Environmental Sampling and Analysis: A Practical Guide

by

Lawrence H. Keith, Ph.D.

Library of Congress Cataloging-in-Publication Data

Keith, Lawrence H., 1938–
Environmental sampling and analysis: a practical guide/Lawrence H. Keith
p. cm.
Includes bibliographical references and index.
ISBN 0-87371-381-8
1. Environmental monitoring. I. Title.
TD193.K45 1991
628.5′028′—dc20 90-26006
CIP

Direct all inquiries to CRC Press, Inc., 2000 Corporate Blvd., N.W., Boca Raton, Florida 33431.

PRINTED IN THE UNITED STATES OF AMERICA
7 8 9 0
Printed on acid-free paper

To my wife, Virginia, who
cheerfully endures these publications.

Dr. Lawrence H. Keith is a Senior Program Manager and Principal Scientist at Radian Corporation in Austin, Texas, A pioneer in environmental sampling and analysis, method development, and handling of hazardous compounds, Dr. Keith has published over a dozen books and presented and published more than one hundred technical articles. Recent publications have involved electronic book formats and expert systems. Dr. Keith serves on numerous government, academic, publishing, and environmental committees and is a past chairman of the ACS Division of Environmental Chemistry. Prior to joining Radian Corporation in 1977, he was a research scientist with the U.S. Environmental Protection Agency.

PREFACE

The detection of a chemical and the setting of standards for regulating its environmental exposure are separate activities; one comprises a strictly technical activity and the other a societal value judgment [McCormack]. McCormack also makes the point that, "unfortunately, there has been too much uncertainty and confusion regarding the real meaning and reliability of detection limits for their optimal use in the world of legislation and regulation. I urge you to work towards a consistent definition and meaningful realization of the limit of detection from a technical point of view, so that it can be accepted by society as a limit above which chemicals will be found (if present) with an acceptable degree of certainty. This means that such limits must fully take into account the real sample characteristics including the effects of interference, contamination, etc. Otherwise, *spurious levels* for false positives or false negatives may lead to regulations that are too restrictive or to pressures that make them too lenient."

The American Chemical Society Committee on Environmental Improvement supports this concept as a logical and responsible course of action. The goal of this guide is to provide a basic understanding of the principles that affect choices made in the many and complex steps involved in environmental sampling and analysis.

Most non-scientists do not appreciate that a billion is one thousand million and do not comprehend the meaning of "one part per billion." Furthermore, most U.S.A. scientists do not appreciate that the word "billion" in the British and German systems is equivalent to "trillion" in the U.S. system, i.e., 10^{12} [Haeberer, 1990].

The ability to detect a minute trace of a contaminant does not necessarily make its removal technically or economically practical.

This is not immediately obvious to the layman. In fact, laymen do not easily comprehend the concepts of very large and very small numbers. As an example, consider the level of one part per billion (1 μg/L) of a pollutant in drinking water. At one part per billion, there are still trillions of molecules of a foreign substance in a glass of water. Another way to express this concept is that "chemically pure" water (99.9999% pure) still contains impurities at a level of one part per million. As McCormack points out, the average citizen is shocked to learn this. This could equate to 100 different substances, each at levels of 10 parts per billion. It is difficult to accurately analyze many substances at the parts per billion level; equipment (both sampling and analytical), reagent impurities, and the presence of other substances at similar or higher levels potentially can contaminate environmental samples.

Previous publications tend to focus on either sampling or analytical activities, but of course, the two procedures are very interdependent. This book, while discussing the principles of each in separate sections, emphasizes the important relationships between sampling and analysis. This is not a book on how to perform environmental sampling and analysis, rather it considers what aspects of sampling and analysis can successfully obtain *data of a known quality*. Notice that the key concept is "data of a known quality"—this means that you should be able to verify that the data produced from an environmental sampling and analysis effort meets your data quality objectives (however stringent or lenient they might be) or that you can recognize when data quality falls short of your objectives. Much environmental analytical data being produced still is of unknown or questionable quality. It is unfortunate, however, that both the producers and users of this data believe that it meets their data quality objective.

It is the purpose of this book to help you to ask questions that will allow you to recognize whether data does or does not meet your own data quality objectives, or if you are planning work, to incorporate decisions to ensure that future data will meet your data quality objectives.

Lawrence H. Keith

The 1990 ACS Joint Board/Council Committee on Environmental Improvement

MEMBERS

Joan B. Berkowitz, Chair
Berkowitz International, Inc.

William Bernaek, Jr.
Indianapolis Center for Advanced Research, Inc.

Michael E. Burns
Michael E. Burns and Associates

Wendall H. Cross
Georgia Institute of Technology

Allan M. Ford
Monsanto Company

A. Wallace Hayes
R.J.R.-Nabisco, Inc.

Fred Hoerger
Dow Chemical Co.

Eugene L. Holt
Levy, Bivona & Cohen

Richard C. Honeycutt
DTRL Inc.

Robert Jamieson
Procter & Gamble Company

Dr. Nathan J. Karch
Karch & Associates, Inc.

Anne R. Keller-Leslie
U.S. EPA

William R. Moomaw
Tufts University

Glenn E. Schweitzer
Soviet and East European Affairs
National Academy of Sciences

Fred Tomboulian
Oakland University

Stephen D. Ziman
Chevron USA, Inc.

ASSOCIATES

John K. Backus
Mobay

Robert L. Jolley
Oak Ridge National Laboratory

William H. Batschelet
Kelly Air Force Base

CONSULTANTS

Geraldine V. Cox
Chemical Manufacturers Association

Lawrence H. Keith
Radian Corporation

Nina I. McClelland
National Sanitation Foundation

SUBCOMMITTEE MEMBERS NOT ON COMMITTEE

David L. Smith
U.S. EPA

John K. Taylor
National Institute of Standards and Testing

AMERICAN CHEMICAL SOCIETY STAFF

Keith Belton
American Chemical Society

Susan M. Turner
American Chemical Society

ACKNOWLEDGMENTS

The author gratefully acknowledges the helpful suggestions provided by the following peer reviewers:

Richard Alabert, Food & Drug Administration; Julian B. Andelman, University of Pittsburgh; Robin Austerman, Shell Development Co.; Michael J. Barcelona, Western Michigan University; Stuart C. Black, U.S. EPA; Gordon Lee Boggs, CA Regional Water Quality Control Board; Olin C. Braids, Blasland, Bouck & Lee; Paul Britton, U.S. EPA; William L. Budde, U.S. EPA; Michael E. Burns, Michael E. Burns & Associates; Joan Bursey, Radian Corp.; John B. Clements, U.S. EPA; William A. Coakley, U.S. EPA; Ursela M. Cowgill, Dow Chemical Company; Geraldine V. Cox, Chemical Manufacturers Association; Lloyd A. Currie, NTIS; Carla Dempsey, Lockheed; William T. Donaldson, U.S. EPA; Aubry E. Dupuy, U.S. EPA; Jim Eichelberger, U.S. EPA; Alan Elzerman, Clemson University; Evan J. Englund, U.S. EPA; Joseph B. Ferrario, U.S. EPA; George T. Flatman, U.S. EPA; Allan M. Ford, Monsanto Company; Lew French, ICF Kaiser Engineers; David Friedman, U.S. EPA; Forest C. Garner, Lockheed; Fred Haeberer, U.S. EPA; David Haile, Monsanto; Robert C. Hanisch, Environmental Quality Consultants, Inc.; Deborah C. Hockman, WMI Environmental Monitoring Labs; Michael T. Homsher, National Sanitation Foundation; William Horwitz, Food & Drug Administration; Andre G. Journel, Stanford University; Cliff J. Kirchmer, Washington State Dept. of Ecology; Shri Kulkarni, Research Triangle Institute; David L. Lewis, Radian Corp.; Robert G. Lewis, U.S. EPA; James P. Lodge, consultant; Eugene P. Meier, U.S. EPA; Marilyn A. Melton, Radian Corp.; Michael J. Messner, Research Triangle Institute; Roger A. Minear, University of Illinois; David J. Munch, U.S. EPA; Lorance H. Newburn, ISCO; Jerry L. Parr, Enseco; Luke A. Petkovsek,

Radian Corp.; Marvin D. Piwoni, Illinois Hazardous Waste Research and Information Center; Thomas H. Pritchett, U.S. EPA; Lloyd P. Provost, Radian Corp.; L. Buck Rogers, University of Georgia; Margaret Rostker, U.S. EPA; Glenn E. Schweitzer, Soviet and East European Affairs, National Research Council, NAS; Michael C. Shepherd, Radian Corp.; Fred L. Shore, Radian Corp.; Guy F. Simes, U.S. EPA; David L. Smith, U.S. EPA; Franklin Smith, Research Triangle Institute; Terry D. Spittler, Cornell University; George Stanko, Shell Development Co.; Martin A. Stapanian, Lockheed; David P. Steele, Environmental Resources Management; Kenneth Stoub, WMI Environmental Monitoring Lab; William A. Telliard, U.S. EPA; Sheri J. Tonn, Pacific Lutheran University; William T. Trotter, Food & Drug Administration; Victor Turoski, James River Research Center; Al W. Verstuyft, Chevron Research Company; John Warren, U.S. EPA; Llewellyn R. Williams, U.S. EPA; Mary M. Walker, Geological and Environmental Services, Inc.; and Stephen D. Ziman, Chevron U.S.A.

CONTENTS

SECTION I. ENVIRONMENTAL SAMPLING

SECTION II. ENVIRONMENTAL ANALYSIS

I

ENVIRONMENTAL SAMPLING

INTRODUCTION TO ENVIRONMENTAL SAMPLING

In 1983 the ACS Committee on Environmental Improvement (CEI) published an article in Analytical Chemistry entitled "Principles of Environmental Analysis." The article contained a section on sampling and its relationship to the analytical results, but it didn't really address this complex subject to the depth that it deserves. In the spring of 1987 CEI sponsored a symposium on "Principles of Environmental Sampling" at the National ACS meeting in Denver, CO. The proceedings of this symposium were published in 1988. Since that time the committee has summarized the most important principles and now supplements them here with an extensive peer review. A condensed version of this summary has been published in *Environmental Science and Technology* [Keith, 1990].

Also in 1990, ACS produced an electronic (computerized) version of "Principles of Environmental Sampling." An advantage of the electronic version is that all words become key words and can be located anywhere they appear in the book. This is in contrast to a manually produced printed index which cannot be sufficiently comprehensive for a subject this complex and having so many key words. Thus, the electronic edition is meant to supplement the printed book as a companion text that is totally searchable. This is a new idea, and publication of the electronic book by ACS software is the first attempt by the society to evaluate the advantages of this concept. These related publications are summarized in Table 1.

Table 1. Related Publications

Date	Reference and Source
1983	"Principles of Environmental Analysis," L.H. Keith, W. Crummett, J. Deegan, Jr., R.A. Libby, J.K. Taylor, G. Weutter, *Anal. Chem. 55*, 2210–2218, 1983.
1987	"Quality Assurance of Chemical Measurements." J.K. Taylor, Lewis Publishers, 121 S. Main Street, Chelsea, MI, 48118. Telephone: 1-800-272-7737.
1988	"Principles of Environmental Sampling," L.H. Keith, ed., ACS Books, Distribution Office, Dept. 258, P.O. Box 57136, West End Station, Washington, D.C. 20037. Telephone: 800-227-5558.
1990	"Principles of Environmental Sampling, Electronic Edition," L.H. Keith, ed., ACS Software, Distribution Office, Dept. 118, P.O. Box 57136, West End Station, Washington, D.C. 20037. Telephone: 800-227-5558.

As with "Principles of Environmental Sampling," the goal of this book is to encourage adequate consideration of the principal variables involved and special techniques needed in planning and carrying out reliable sampling of environmental matrices. Specific needs will dictate which approaches are incorporated in sampling plans and which are rejected. This book emphasizes broad principles, while "Principles of Environmental Sampling" contains many more details.

There are several themes common throughout Section I. First, samples that are not representative of the population of interest are of little use. Second, poor sample collection procedures yield unrepresentative samples and contribute to the uncertainty of the analytical results. Furthermore, sampling errors and analytical errors occur independently of each other, so sampling related errors cannot be accounted for by laboratory blanks or control samples. If the sampling procedures are carefully and thoroughly documented, sometimes they can identify poor sampling protocol errors; however, the errors can rarely be corrected without resampling and analysis. The final theme is that contamination is a common source of error in all types of environmental measurements. Most sampling and analytical schemes present numerous opportunities for sample contamination from a variety of sources.

Sample contamination may come from artifacts in sampling containers. Artifacts may also be introduced during sample collection, preservation, handling, storage, or transport to the laboratory. After samples arrive at a laboratory, additional opportunities for contamination arise during storage, in the preparation and handling process, or in the analytical process itself. Thus, major sections of this book deal with appropriate quality assurance (QA) and quality control (QC) procedures which will ensure that representative samples are collected, that introduction of contaminants during the sampling phase is minimized, and that contamination is identifiable when it does occur.

Separate chapters are devoted to special problems associated with sampling different matrices—water, air, biota, solids, sludges, and liquid wastes.

CHAPTER 1

Planning and Sampling Protocols

The objective in collecting samples for analysis is to obtain a small and informative portion of the population being investigated. Usually representative samples are sought, i.e., samples that can be expected to adequately reflect the properties of interest of the population being sampled. Sometimes targeted, or nonrepresentative, samples are needed. An example might be a particular site near an industrial outfall suspected of polluting a river. However, sampled aliquots usually need to be representative of that site at the time the samples were taken to be useful. If the samples, individually and collectively, cannot provide the required information, they are seldom worth the time and expense of analysis. Therefore, planning for informative sampling must be an integral part of any study. Often, confidence in the data quality depends on the number of samples and the location and time of sampling [Kulkarni].

Figure 1 provides a convenient checklist of the subjects that should be considered when planning sampling episodes. These topics are discussed in detail in the following subsections.

DATA QUALITY OBJECTIVES AFFECT ALL ASPECTS OF PLANNING

Data quality objectives (DQOs) are statements that provide the critical definitions of confidence required in drawing conclusions from the entire project data. These objectives determine the degree

What are your data quality objectives (DQOs)?
- What will you do if your DQOs are not met (i.e., resample or revise DQOs?)

Do program objectives need exploratory, monitoring, or both sampling types?

Have arrangements been made to obtain samples from the sites?
- Have alternate plans been prepared in case not all sites can be sampled?

Is specialized sampling equipment needed and/or available?

Are samplers experienced in the type of sampling required/available?

Have all analytes been listed?
- Has the level of detection (LOD) for each been spacified?
- Have methods been specified for each analyte?
- What sample sies are needed based on method and desired LOD?

List specific GLP, federal, state, or method QA/QC protocols required.
- Are there percentages or required numbers and types of QC samples?
- Are ther specific instrument tuning or other special requirements?

What type of sampling approach will be used?
- Random, systematic, judgmental, or a combination of these?
- Will the type of sampling meet your DQOs?

What type of data analysis methods will be used?
- Geostatistical, control charts, hypothesis testing, etc.?
- Will the data analysis methods meet your DQOs?
- Is the sampling approach compatible with data analysis methods?

How many samples are needed?
- How many sample sites are there?
- How many methods were specified?
- How many test samples are needed for each method?
- How many control site samples are needed?
- What types of QC samples are needed?
  - Will the QC sample types meet your DQOs?
- How many of each type of QC samples are needed?
  - Are these QC samples sufficient to meet your DQOs?
- How many exploratory samples are needed?
- How many supplementary samples will be taken?

Samples = Test + Control + QC + Exploratory + Supplementary
- Test Samples = Methods x Sample Sites x Samples per Site
- Control Samples = Methods x Sample Sites x Samples per Site
- QC Samples = Methods x Type of QC Sample x % needed to meet DQOs
- Exploratory Samples = (Test Samples + Control Samples) x 5 to 15%
- Supplementary Samples = (Test Samples + Control Samples) x 5 to 15%

Figure 1. Sampling plan checklist.

of total variability (uncertainty or error) that can be tolerated in the data [Meier]. These limits of variability must be incorporated into the sampling and analysis plan and achieved with detailed sampling and analysis protocols. DQOs differ from measurement quality objectives (such as precision and accuracy) in that they are limits for the overall uncertainty of results, while the latter are only limits for the uncertainty of specific measurements.

Data Quality Objectives may be qualitative or quantitative. Qualitative DQOs are specific descriptions of actions that are to be taken if an answer does not meet the DQO. They contain no quantitative terms but reflect general decisions that must be made.

For example, what action is going to be taken if QC samples are found to contain significant amounts of contamination? Possible answers include (1) discarding the data and doing without it, (2) tracking down the problem and resampling, or (3) only using data with analyte concentrations above the background contamination level.

Quantitative DQOs contain specific quantitative terms such as standard deviations, relative standard deviations, percent recovery, relative percent difference, and concentration.

For example, if the desired limit of detection or a reliable detection level (discussed in Section II) cannot be met, what actions will be taken? Possible answers include (1) accept the higher detection level, (2) composite several samples to obtain a larger sample, (3) try a different sample preparation (cleanup) technique, (4) try a different analytical method, (5) resample the same site and take larger samples, (6) resample at a different site, or (7) resample using a different sampling technique at the same or a different site.

Often, desired data quality objectives must be balanced against the cost of a sampling and analysis project, and more realistic objectives must be adopted with concurrence of the data users. Three factors that most influence the cost of sampling are (1) site location and accessibility to sampling points, (2) the numbers, kind, complexity, and size of samples to be collected, and (3) the frequency of sampling. The extent to which these factors will influence cost depends on particular aspects of each sampling project. Case studies of previous efforts can be important in developing realistic data quality objectives.

BASIC CONSIDERATIONS OF PLANNING

Results of environmental monitoring studies, especially if they are published, are frequently used for purposes beyond those of the original investigation. In addition, sometimes data are obtained for multiple uses; thus, a sampling plan may, by necessity and design, become a compromise among the demands of various data quality objectives. This situation occurs frequently and must be recognized and documented to the fullest extent possible. Ideally, a compromise on data quality should be avoided, and sampling designs should be based on the most stringent data quality objective or by the most stringent anticipated use. Realistically, however, this is not always necessary. Using the most stringent and expensive data quality requirements instead of thoroughly understanding the data quality objectives (and perhaps providing good advice to data users) can needlessly expend resources and raise costs of both sampling and analysis [Homsher].

Sometimes sampling and analytical activities are terminated prior to the completion of the full sampling program for financial, political, or other reasons. Therefore, a monitoring program should be designed in stages, and each stage should provide meaningful results.

The first step in planning a sampling activity is to define clearly the objective. Objectives of environmental sampling are broadly divided into exploratory (surveillance) or monitoring (assessment) goals. Exploratory sampling is designed to provide preliminary information about the site or material being analyzed. Monitoring, on the other hand, usually is intended to provide information on the variation of specific analyte concentrations over a particular period of time or within a specific geographic area; monitoring is used for regulatory (enforcement) and nonregulatory purposes. A sampling plan for monitoring usually is more effective if it is preceded by exploratory sampling or if there is historical data on the analytes of interest at the sampling site. Exploratory sampling can help establish the chemical species of concern and the range of their concentrations and variability. This optimizes the selection of sampling equipment, analytical methodology, numbers of samples, and sampling protocol and establishes meaningful QC criteria.

Because different analytical techniques are used for different species, sufficiently large samples must be taken frequently for multiple analyses. Also, since analytical techniques are not well developed for some analytes in complex matrices, large samples provide laboratories the opportunity to analyze replicate samples or reanalyze samples when the data is suspect. However, disadvantages of large samples include additional costs of storage space, materials, and larger sample disposal costs.

Control sites are important in understanding the significance of monitoring data. Select sites that have common characteristics with the affected areas except for the pollution source, and take adequate numbers of control samples to permit meaningful comparisons. This is discussed in detail later.

Often, preliminary sampling (screening) is desired to help delineate the extent of contamination and the variations in contaminant levels within the affected area. This preliminary sampling may involve 10 to 15% of the overall monitoring effort. It requires an additional step of preliminary data analysis before the remaining samples are collected [Journel]. When conducting preliminary sampling it is important that both the sampling and subsequent analyses, or preliminary work, be performed under the same sampling, analytical, and QA/QC protocols as those developed for the main body of test samples. Otherwise, the preliminary sampling may produce invalid data and false conclusions [Dupuy, 1989].

Frequently, supplementary sampling (resampling) also is desirable; it confirms particularly critical findings and clarifies uncertainties that were discovered during the monitoring program. This supplementary sampling may also involve 10 to 15% of the monitoring effort.

A sampling program design also must consider the quality of the data needed, i.e., the degree to which total error must be controlled to achieve the required level of confidence. The data collection planning process should provide a logical, objective, and quantitative balance between the time and resources available for collecting the data and the data quality based on its intended use. A pre-established budget should not be the sole constraint on the design of a data collection program [Haeberer, 1989]. If the desired data quality and representativeness cannot be obtained within budget constraints then the decision to increase the budget and resources or to accept reduced data quality or representativeness must be made [Meier].

One of the most important aspects of the data collection planning process is the joint involvement of the data users and the samplers and analysts. The initial and continued involvement, and the perspective of each, is critical to defining data quality and quantity requirements.

The planner must also determine whether a study should be conducted according to regional, state, or federal protocols (such as FDA, EPA, TSCA, CERCLA, and FIFRA) or other Good Laboratory Practice (GLP) regulations, since these mandate specific, documented practices that affect sampling and analysis. Studies conducted according to these regulations require adherence to strict protocols; otherwise the resultant data may be unacceptable.

The choice of a data analysis method is an important decision that also should be made in the planning stage. It must facilitate, and be facilitated by, the goals, DQOs, and experimental design. Both analyses and sampling approaches (judgmental, random, and systematic) require prior information to meet data quality objectives. Any random variable method of data analysis (such as hypothesis testing, estimation interval, tolerance interval, control charts, etc.) requires random sampling. The number of samples for random variable methodology must be determined by the population variance and the desired size of a "significant change" in the test parameter.

Systematic sampling is preferable for geostatistical data analysis, but random or even judgmental sampling may achieve greater accuracy within specific local areas; random or judgmental sampling may also meet other specific purposes. Geostatistical data analysis accounts for the time and space dependence of data, and it is usually used to produce site maps (with qualification of interpolation errors) showing analyte locations and concentrations. Geostatistical data analysis is an iterative process that determines patterns of space/time correlations. The data analysis tool used to accomplish this is variogram distance plots. These plots of variance in analyte concentrations come from paired sample measurements and are a function of the distance between the samples. An interpolation technique called kriging then allows sample weighting according to their variogram distances and corresponding ranges of correlation. Prior information is extremely important with any monitoring design [Flatman and Journel]. However, judgmental sampling, in the sense of purposely trying to obtain the highest concentration of analytes possible in test samples, is not a geostatistical type of sampling [Englund].

The selection of an appropriate sampling technique is frequently made on a case by case basis. Even when a technique is prescribed in regulations, technical constraints imposed in the field may require minor modifications. Therefore, complete documentation of the sampling technique in the sampling protocol, together with assessments of the uncertainties that may be associated with it, is important. Field comparisons of alternative sampling techniques, prior to the final selection, are often helpful in selecting an approach with the fewest uncertainties for the particular situation. However, such comparisons generally provide only relative data since the "true values" of the analytes are usually unknown at this stage [Barcelona, 1989].

Two basic sampling decisions, that must be resolved during the planning stage and documented in the sampling protocol, are the types and numbers of QC samples to take [Keith, 1988]. The answers directly depend on the nature of the errors to be assessed (both systematic and random) and the accuracy desired in their assessment. Additional considerations include the contribution of sampling error relative to total error, the relative cost of sampling and analysis, and the sensitivity and selectivity of the analytical method in relation to the concentration of the analytes [Messner].

For sampling activities undertaken in accordance with permit requirements, the sampling frequency may be predetermined. However, when sampling is undertaken for other purposes, the sampling design, including the numbers and sizes of samples and the order in which they are taken, must be determined by considering the variability of the target analytes in the samples, the analytical methodology, constraints on access to the sampling area, costs, and the DQOs. Any questions that a study intends to answer must be asked prior to establishing the sampling protocol. The protocol should then be evaluated to insure that the samples can provide the answers sought [Dupuy, 1989].

Careful documentation during sampling plan preparation will help eliminate misunderstandings later in the sampling effort. This documentation should include the objectives of the sampling, a description of the location, timing, and character of the samples, preservation precautions, a sample identification, chain of custody records, and an indication of the analyses that are to be made. This information, which is gathered and documented during the planning process, is also used in the sampling protocol.

GOOD SAMPLING PROTOCOLS DOCUMENT PLANNING

Sampling protocols are written descriptions of the detailed procedures to be followed in the collection, packaging, labeling, preservation, transportation, storage, and documentation of the samples. The more specific a sampling protocol is, the less chance there will be for errors or erroneous assumptions. Figure 2 provides a convenient checklist of topics that should be considered when preparing sampling protocols.

What observations at sampling sites are to be recorded?

Has information concerning DQOs, analytical methods, LODs, etc., been included?

Have instructions for modifying protocols in case of problems been specified?

Has a list of all sampling equipment been prepared?
- Does it include all sampling devices?
- Does it include all sampling containers?
  - Are the container compositions consistent with analytes?
  - Are the container sizes consistent with the amount of samples needed?
- Does it include all preservation materials/chemicals?
- Does it include materials for cleaning the equipment?
- Does it include labels, tape, waterproof pens, and packaging materials?
- Does it include chain of custody forms and sample seals?
- Does it include chemical protective clothing or other safety equipment?

Are there instructions for cleaning equipment before and after sampling?
- Are instructions for equipment calibration and/or use included?
- Are instructions for cleaning or handling sample containers included?

Have instructions for each type of sample collection been prepared?
- Are numbers of samples and sample sizes designated for each type?
- Are any special sampling times or conditions needed?
- Are numbers, types, and sizes of all QC samples included?
- Are numbers, types, and sizes of exploratory and supplementary samples included?
- Are instructions for compositing samples needed?
- Are instructions for field preparations or measurements included?

Have instructions for completing sample labels been included?

Have instructions for preserving each type of sample been included?
- Do they include maximum holding times of samples?

Have instructions for packaging, transport, and storage been included?

Have instructions for chain of custody procedures been included?

Have safety plans been included?

Figure 2. Sampling protocols checklist.

Sampling protocols should contain written instructions for all sampling activities including observations at the sample site, and field documentation of sampling techniques. In addition, most protocols should have a statistical design to prove that the samples represent the matrix to be evaluated [Meier]. This statistical design should, together with the sample preparation and analytical methods, respond to the DQOs and data collection objectives of the study. Always remember that statistical equations are tools to aid common sense and not substitute it.

In preparing a statistical sampling design, the variability of the sampling and analytical process is frequently considered constant (homoscedastic) over time, space, and other variables [Liggett]. The more general model recognizes that variability is not constant (heteroscedastic); it depends on sampling and analysis time and location and on analyte concentration.

The overall sampling protocol must identify sampling locations and include all of the equipment and information needed for sampling: the types, numbers, and sizes of containers, labels, field logs, types of sampling devices, numbers and types of blanks, sample splits and spikes, the sample volume, any composite sample specifics, preservation instructions for each sample type, chain of custody procedures, transportation plans, any field preparations (such as filtering or pH adjustments), any field measurements (such as pH, dissolved oxygen, etc.), and the report format [Bone]. Also, it should identify those physical, meteorological, and hydrological variables that should be recorded or measured at the time of sampling [Barcelona, 1988]. In addition, information concerning the analytical methods to be used, minimum sample volumes, desired minimum levels of quantitation, and analytical bias and precision limits may help sampling personnel make better decisions when unforeseen circumstances require changes to the sampling protocol.

Selecting analytical methods is an integral part of the sample planning process and can strongly influence the sampling protocol. For example, the sensitivity of an analytical method directly influences the volume of sample necessary to measure analytes at specified minimum detection (or quantification) levels. The analytical method may also affect the selection of storage containers and preservation techniques [Taylor, 1988]. Whether the sampling and analytical protocols are combined in a single document usually depends on whether the same organization is responsible for both sampling and analysis. When two different organizations are responsible for sampling and analysis, separate protocols are usually necessary [Horwitz].

BASIC SAMPLING APPROACHES

The sampling protocol must be carried out by personnel trained and experienced in the specified sampling techniques and procedures. Sampling approaches vary greatly, depending on the objectives of the study and the complexity of the sampling site. There are three primary sampling approaches: random, stratified, and judgmental. There are also three primary combinations of each of these: stratified (judgmental) random, systematic random, and systematic judgmental. And there are further variations that can be found among the three primary approaches and the three combinations of them. For example, the systematic grid may be square or triangular; samples may be taken at the nodes of the grid, at the center of the spaces defined by a grid, or randomly within the spaces defined by a grid. Table 2 summarizes the differences among the three primary approaches, and Figure 3 illustrates them. Figure 4 illustrates the correlation between the approaches and their combinations vs the relative number of samples needed and the relative amount of bias incorporated into each.

Samples collected purely for research studies may make use of prior knowledge about the site in order to obtain as much useful data as possible; however, sampling for legal purposes often requires absolutely random samples so that all personal bias is removed. An advantage of random sampling is the simplicity of assumptions about the sample population. A disadvantage of random sampling is that it is usually more expensive; sometimes money is essentially substituted for good judgment if random sampling is strictly observed [Borgman].

When prior knowledge is used to select sampling sites, the reasons for each decision must be carefully documented. Introducing some

Table 2. Primary Sampling Approaches

Approach	Relative Number of Samples	Relative Bias	Basis of Selecting Sampling Sites
Judgmental	Smallest	Largest	Prior history, visual assessment of technical judgment
Systematic	Larger	Smaller	Consistent grid or pattern
Random	Largest	Smallest	Simple random selection

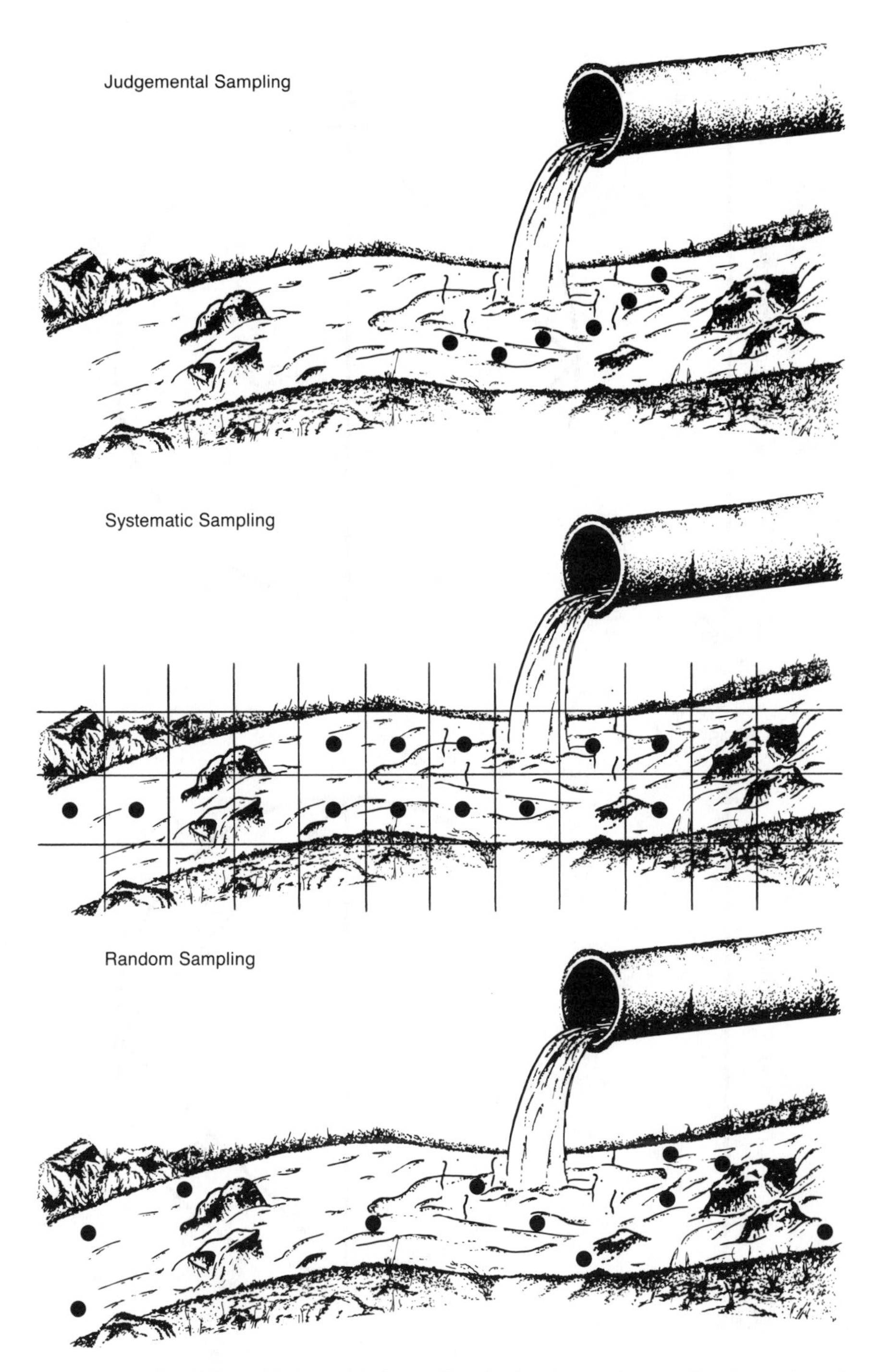

Figure 3. Examples of judgemental sampling (top), systematic sampling (center), and random sampling approaches (bottom).

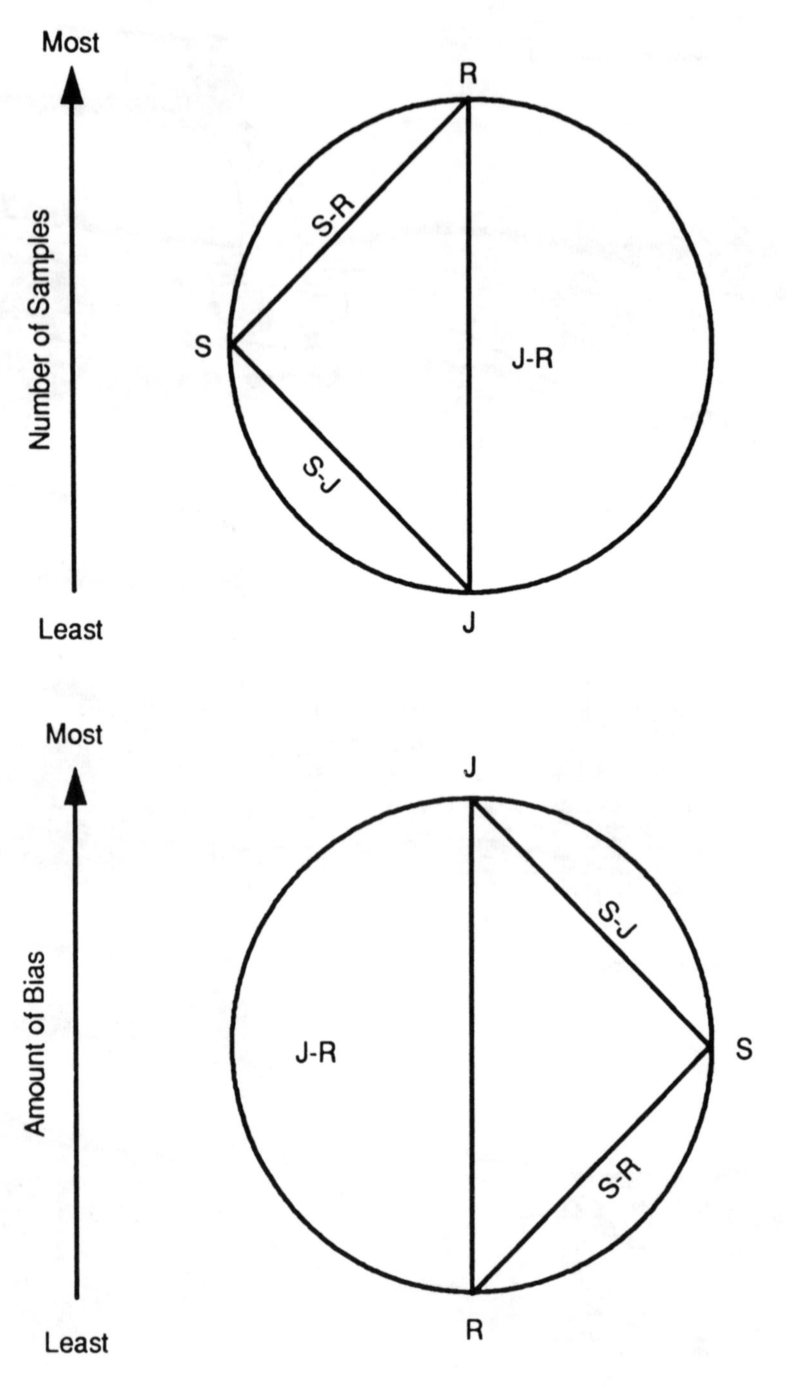

Figure 4. The general relationship of random (R), systematic (S), and judgmental (J) sampling approaches to the relative number of samples needed (top) and the relative amounts of bias (bottom).

random sampling is usually desirable and possible without causing unacceptable, additional costs. The greatest danger in substituting judgmental sampling for random sampling is the difficulty of knowing (or proving) whether or not various assumptions are acceptable for a particular application [Borgman].

Since most analytes show some type of space/time (temporal) dependence, some kind of stratified random or search sampling is usually most efficient [Journel, 1989]. Sample sites that are more heterogeneous (nonuniform) than others, or that cover relatively large areas or volumes, are more difficult to sample, and they usually require more samples to achieve a desired level of information.

Often a combination of judgmental, systematic, or random sampling is the most feasible approach; however, the sampling scheme should be sufficiently flexible to permit adjustments during field activities. Problems such as a lack of access to preselected sampling sites or unanticipated subsurface formations may necessitate adjustments. All deviations from the sampling protocol must be fully documented.

PLANNING INCLUDES ASSESSING MIGRATION OF POLLUTANTS

Most environmental pollutants do not remain stationary. If they are in a water, air, soil, sludge, solid, or liquid matrix they are almost certain to migrate. The physical characteristics of each matrix, meteorological conditions, the amount of pollutant present, the rate of release into the environment, the source of release and human intervention all affect the pathway and rate of migration.

The most common transport mechanisms for environmental pollutants are wind, rain, surface water, groundwater, and human intervention (wastewater pipes, drainage ditches, roads, railways, etc.). In addition to transport mechanisms, physical and biological influences may also affect migration of pollutants. Physical influences include topographical features (valleys, mountains, slopes, lakes, rivers, etc.) and geological features (aquifers, soil composition, mineral composition, etc.). These physical influences can either aid or impede chemical migration. Biological influences usually consist of food pathways. Bioaccumulation of environmental pollutants, from low concentrations in water, air, and soil to increasingly higher concentrations through the food pathways of plants and animals, is well documented and must be carefully considered when sampling biota.

Often the main objective of an environmental study will be to determine how far pollutants have migrated from their source and to measure their concentrations at various distances from their source or some other point of interest. Regardless of the objective of a study, migration is always an important issue when obtaining blanks from nearby control sites. Analytes of interest migrating into the control site blanks, when the blanks are supposed to contain only background amounts of those analytes, will bias the test results with low values when high background levels are subtracted from the test sample data. This is discussed in more detail in the Quality Assurance and Quality Control Section.

PLANNING INCLUDES SELECTING APPROPRIATE SAMPLING DEVICES

Sample collection must not significantly disturb the environment, or the site will be changed and the analytical results will be biased. For example, taking grab samples for hydrocarbon analysis from a motor boat in a lake is likely to bias the samples with hydrocarbons from the motor. The methods and materials used to collect, store, and transport samples should be considered carefully. Sample collection involves contact with a sampling device and its materials of construction; therefore, the performance of sampling and storage equipment, the degree of systematic error, and sampling precision must be evaluated under specified conditions.

Contamination by sampling devices and materials can contribute relatively large errors in comparison to analytical procedures, especially when the analytes of interest are at low concentrations. If the sample contacts materials that are reactive or sorptive, or that leach either the analytes or analytical interferents, they greatly bias the sample source. The components of any sampling device, e.g., tubing, gaskets, metal or plastic components, and sorptive materials, must be carefully evaluated and checked for analyte interaction [Kent].

The selection of a sampling device and the sampling protocol should be based on the most labile analytes to be measured. Labile in this context refers to the probability that the analyte concentration will change prior to analysis. Important considerations are the analyte reactivity (with light, heat, air, water, biological organisms, metals, and other chemicals), volatility, space/time variability, and irreversible sorption potential by the sampling device and/or storage container. Often replicate samples (multiple samples taken under comparable conditions) must be collected and handled differently (in terms of preservation or sample preparation) in monitoring for different analytes. If field QC samples are included in the sampling protocol they also should be collected in sufficient numbers so that all the types of samples taken are represented.

COMPOSITE SAMPLING MUST BE PLANNED CAREFULLY

In typical monitoring programs each individual test sample is analyzed; however, composite sampling (combining portions of multiple samples) can provide advantages in the following situations:

- When samples taken from various locations or populations are analyzed to determine if the component of interest is present
- When aliquots of extracts from various samples composited for analysis are analyzed to determine whether the component of interest is present
- When representativeness of samples taken from a single site or population needs to be improved by reducing intersample variance effects
- When representativeness of random samples, removed from a potentially heterogeneous matrix, needs to be ensured by reducing the effect of variance between aliquots
- When a necessarily limited size of the material available for analysis, such as blood specimens, needs to be increased to achieve analytical performance goals [Garner, et al.]
- When estimating the frequency of a trait, such as the HIV virus, to reduce the cost and/or mean square error of the estimate while, at the same time, preserving the confidentiality of the sampled individuals [Stapanian and Garner].

Composite sampling is often used to reduce the cost of analyzing a large number of samples. Experimental costs are substantially reduced when the frequency of individual samples containing the analytes of interest is low. In such experiments, individual sample aliquots are combined into composites, and each composite is analyzed. If any composite analysis result exceeds the ratio of a specified criterion level to the number of sample aliquots composited, then each individual sample in that composite is analyzed to determine which possesses the analyte above the criterion level. To avoid false negatives, the number of sample aliquots combined into each composite must not exceed the ratio of the criterion level to the limit of detection. Composite sampling provides the added benefit of reducing the number of false positives. A false positive will occur in the unlikely event that both the composite and one or more individual samples falsely test positive [Stapanian and Garner].

Composite sampling may also reduce intersample variance due to the heterogeneity of the sampled material. When estimating the frequency of an analyte occurrence, the frequency of composites that test positive (composite samples where the analytes of interest are above the detection level) is related to the frequency of individual samples that test positive. Thus the frequency of the analyte occurrence in the sample population may be estimated as a function of the observed frequency of positives for the composites. Because analytical costs are reduced, more samples may be taken. This approach can lead to a frequency estimator of the analyte with less mean square error than the traditional approach. This is generally true when sampling costs are much less than analytical costs and the frequency of occurrence is low. No samples need to be reanalyzed with this approach. As a result, the anonymity of the sampled individuals is preserved. This may lead to a less biased sample and appeal to people who might be reluctant to participate due to potential implications of a positive result on an individual sample.

Composite sampling may also increase the amount of sample material available for analysis. This would allow for some types of analyses for which individual samples do not provide sufficient material. Suppose that a particular analytical method required 2 L of human blood to achieve the necessary data quality (detection limit). Although an individual human cannot be safely sampled to this extent, a composite made of blood samples from all members of a family or household could achieve the DQO. The result of the analysis would apply to the household in general and not necessarily to any individual. This same philosophy can be applied to any identifiable group, such as an industrial process, from which samples can be taken [Stapanian and Garner].

Compositing also has limitations:

- When composite sampling is used, interactions among analytes or organisms must be considered carefully. Take care to ensure that analytes or organisms from different samples will neither be mutually destructive nor create analytical interferences. If corrective action is not taken when such problems are suspected, composite sampling that is not representative of any of the original samples may result in a test material [Stapanian and Garner].

- When the objective of the monitoring program is a preliminary evaluation or classification, compositing may dilute the analyte to a level below the detection limit, producing a false negative.

- If sampling costs are greater than analytical costs, analyzing each sample individually may be more cost effective.

- When considering multiple analytes in a composite, information regarding analyte relationships in individual samples will be lost.

- If compositing reduces the number of samples collected below the required statistical need of the DQOs, then those objectives will be compromised.

However, if all of these considerations are evaluated during the planning stage, they can be documented and accommodated in the sampling protocol.

SAFETY PLANS REDUCE CHEMICAL EXPOSURE AND LIABILITY

Safety must always be considered in the development of any sampling plan. In the U.S., OSHA requires specific training for sampling at hazardous waste sites [Walker]. Proper planning and execution of safety protocols help protect employees from accidents and needless exposure to hazardous or potentially hazardous chemicals. Documenting safety protocols may also be important in cases of litigation and worker compensation issues as the result of accidents or accidental chemical exposure during sampling episodes. Documentation should include requirements for hard hats, safety boots, safety glasses, respirators, self contained breathing air, gloves, and hazardous materials suits if any of these are needed. In addition, personal exposure monitoring and/or monitoring ambient air concentrations of some chemicals may be necessary to meet safety regulations.

Potential exposure of personnel to hazardous chemicals that can permeate their chemical protective clothing (CPC) causes concern whenever neat chemicals or chemicals in high concentrations, e.g., from some landfills and wastewater streams, are to be sampled. A standard test method (Method F739–85) for measuring chemical permeation through CPC has been established by the American Society for Testing and Materials (ASTM). Many manufacturers of CPC routinely report breakthrough times and permeation rates of their products with various chemicals.

Beware of Material Safety Data Sheets (MSDS) that contain vague, unspecific references to CPC selection. Many different polymers are used in CPC manufacture. Furthermore, variations in the manufacturing process can significantly affect the chemical permeation properties of a garment. Therefore, selection of manufacturer and product model can be of great importance when selecting CPC. Sometimes, many different brochures must be reviewed before a confident selection can be made. Alternatively, "Chemical Protective Clothing Performance Index" [Forsberg] with "GlovES+" is available from ACS Software (see Table 1), leading CPC safety suppliers, and the publisher (Instant Reference Sources, Inc., Austin, TX, (512) 345-5267). These PC computer programs contain comprehensive sources of chemical permeation data by manufacturer, polymer type, and chemical that can be searched; reports that document the search results are also easily generated. Previous editions of Forsberg's data were essentially restricted to IBM and compatible PCs, but in 1990 a new version introduced "HyperCPC Stacks," a HyperCard version that is compatible with Macintosh Plus, SE, and portable MAC type computers. "HyperCPC Stacks" can also access the U.S. Coast Guard "Cameo™ II Navigator" for emergency response situations.

Not all sampling episodes will need special safety consideration, but where they are needed, documented and well designed safety protocols are important adjuncts to sampling protocols. Obviously, if there are special sampling hazards that may affect laboratory personnel, these should be documented and provided to the analysts. In addition to providing this information in written documents, precautionary warnings should also be placed on sample containers; usually there are spaces for comments on sample labels.

When no special safety protocols or equipment are necessary, a statement to that effect should be included in the sampling plans. This will document that safety considerations were not overlooked but were duly considered, and that no special safety equipment or plans were necessary based on the information at hand.

CHAPTER 2

Quality Assurance and Quality Control

The planning-sampling-analysis-reporting chain is composed of many links and the uncertainty of the final result is a function of the uncertainties of each step. There are errors in all parts of this chain, and the goal of quality assurance is to be able to identify, measure, and control these errors. The goal of quality improvement is to minimize or correct for individual errors and their cumulative effect.

Laboratory oriented quality control measures can account only for errors that occur after sample collection; therefore, sampling procedures must have their own QA and QC protocols. The entire sequence of sample gathering, preservation, storage, and shipment must be evaluated to measure and minimize systematic sources of error and ubiquitous sources of random error; these sources cannot be measured using laboratory oriented QA and QC protocols.

The two parameters most often used to assess measurement quality objectives are bias and precision. *Bias* is defined as a systematic deviation (error) in data. *Precision* is defined as random variation in data. One objective of any sampling quality assurance program is to provide the type and number of quality control samples necessary to control and minimize the effects of bias and precision in the sampling effort.

It is important that the design of the intended study incorporates enough samples to meet the statistical and representativeness objectives. A certain number of extra samples should be programmed into the sampling design to allow for a loss of sample data due to breakage, lab mishaps, and analytical, out of control situations. This will ensure that completeness objectives can be obtained [Dupuy, 1989].

SAMPLING CONSIDERATIONS

Sampling operations can contribute to both systematic and random error. These components are interactive, and they are evaluated using appropriate statistical methods. Some aspects of sampling may cause large variations and yet cannot be treated strictly by statistics, i.e., by reducing or controlling systematic errors. Examples include the proper selection of sampling devices, storage containers, and preservation protocols. Sometimes, the selection of the analytical method and even the selection of the sampling sites fall within this category.

Sorption or reaction bias from container materials, sampling equipment, or material-transfer lines can totally invalidate data generated from the samples. Deterioration from biologically mediated reactions, atmospheric contact, temperature instability, radiation, and interactions with other analytes can also cause serious biases. If any of the analytes of interest are subject to deterioration from these effects, the sampling protocol must specify preservation conditions that will prevent or minimize their deterioration.

Sometimes field manipulations are necessary, but they should be planned and recorded carefully and kept to a minimum because they are prone to introducing bias. Material transfers, filtrations, physical measurements, aliquot preparation, and spiking are all potential sources of bias from contamination.

Field QC samples, e.g., blanks, spikes, and replicates, should be handled exactly the same way as the environmental samples. This includes using identical sampling devices, sampling protocol, storage containers, shipping procedures, and preservation techniques. Only field QC samples can provide the basis from which sampling bias (usually contamination) can be estimated or sampling precision calculated.

DEFINING BLANK SAMPLE REQUIREMENTS

Blanks are defined as matrices that have negligible or unmeasureable amounts of the substance of interest [D.L. Lewis, 1988]. The needs for blanks and controls are influenced by the DQOs. Wherever a possibility exists for introducing extraneous material into a collection, treatment, or analytical procedure, a blank should be devised to detect and measure the extraneous material. Blanks commonly used for environmental sampling include field, trip, matrix, equipment, and material blanks. These are usually supplemented at the laboratory with preparation solvent, reagent, instrument, and method blanks to measure analytical bias; these are discussed in Section II.

Field blanks

Field blanks are samples of analyte-free media similar to the sample matrix. They are transferred from one vessel to another or exposed to the sampling environment at the sampling site [Parr, et al., 1989]. They measure incidental or accidental sample contamination during the whole process (sampling, transport, sample preparation, and analysis). Capped and cleaned containers are taken to the sample collection site. Usually each sampling team should collect one field blank a day per collection apparatus; the field blank matrix should be comparable to the sample of interest [Black]. Field blank water samples consist of triple distilled water that is carried to the sampling site and exposed to the air there so that any contamination from the air can be measured and accounted for [Cowgill, 1989].

Trip Blanks

Trip blanks (or *transport blanks*) are test samples of analyte-free media taken from the laboratory to the sampling site and returned to the laboratory unopened [Parr, 1989]. They are used to measure cross-contamination from the container and preservative during transport, field handling, and storage. Usually at least one trip blank should be collected per day per type of sample [Black]. EPA has shown that cross-contamination only occurs with volatile organics (VOAs) [Coakley].

Background Samples

Background samples (or *matrix blanks* or *control samples*) are samples of the media similar to the test sample matrix and are taken near to the time and place where the analytes of interest may exist at background levels. They measure the background presence of analytes of interest in the same matrix from which the samples are obtained (for example, water, sludge, activated carbon, sandy loam soil, tissues, etc.). Usually the frequency of their analysis should be equivalent to that of the reagent blanks [Black]. The more complex a matrix is, the more important matrix blanks become [Burns].

Equipment Blanks

Equipment blanks (or *rinsate blanks*) are samples of analyte-free media that have been used to rinse the sampling equipment. They document adequate decontamination of the sampling equipment after its use. These blanks are collected after equipment decontamination and prior to resampling [Parr, et al., 1989].

Material Blanks

Material blanks are samples of construction materials such as those used in groundwater wells, pump and flow testing, etc. They document the decontamination (or measure artifacts) from use of these materials [Parr, et al., 1989].

USING BLANKS

Sample analysis is often expensive. Sometimes it is prudent to collect a full suite of blanks but only analyze the field blanks. If the field blanks indicate no problems, the other blanks may be discarded or stored as necessary. If a problem is discovered, the individual blanks can be analyzed to determine its source [Black]. Resampling will still likely be necessary. One caveat to this philosophy is the possibility of exceeding the prescribed holding times of the other blanks, rendering them invalid.

When an extraction device (such as an air filter, charcoal, an ion exchanger, or Tenax®) is used instead of direct sample collection, the field, storage, and transport blanks must include the corresponding extraction device. For example, an unused air filter should be unsealed when air filters are used for sampling, and subsequently treated as if it had been used for sampling [Black]. Blank solid sorbents such as Tenax and charcoal, on the other hand, should remain sealed until they are opened for analysis as matrix blanks. This is because they will immediately begin to sorb volatile compounds from the surrounding air [Clements, 1989].

Whether blank data are used for ongoing control or for retrospective assessment, control charts (plots of the data from blanks vs time) provide the most effective mechanism for interpreting blank analysis results [D.L. Lewis, 1988].

BACKGROUND (CONTROL) SAMPLE SELECTION

There are basically two types of controls: (1) those used to determine whether or not an analytical procedure is in statistical control, and (2) those used to determine whether or not an analyte of interest is present in a studied population but not in a similar control population [Black]. The analytical control standards (check standards and laboratory control standards) are discussed in Section II. Sampling control standards, used to determine whether an analyte of interest is really present in the test samples under study, consist of two basic types: background (control site) and field spike samples. Control sites are further differentiated as "local control sites" and "area control sites."

Background Samples

Background samples (or *control site* or *matrix samples*) are samples taken near the time and place of the sample of interest; they were discussed earlier. Background samples demonstrate whether the site is contaminated or truly different from the norm [Meier]. Some sort of background sample is always necessary for a valid scientific comparison of samples suspected of containing environmental contaminants with samples containing no (below detectable or measurable levels) or acceptably low levels of contaminants. Unless background samples are collected and analyzed under the same conditions as the samples of interest, the presence and/or concentration levels of the analytes of interest and the effects of the matrix on their analysis can never be known or estimated with any acceptable degree of certainty. Background samples of each significantly different matrix must always be collected when different types of matrices are involved.

Examples include various types of water, sediments, and soils in or near a sampling site area. Background air samples would include upwind air samples and perhaps different height samples. The only logical exception to collecting background samples is when drums or containers of materials are involved, as in a landfill; however, if the chemicals are suspected of polluting the land, water, or air around them, appropriate background samples of those matrices must be taken for analysis.

Local control sites are usually adjacent or very near the test sample sites. In selecting and working with local control sites the following principles apply:

- Local control sites generally should be upwind or upstream of the sampling site.
- When possible, local control site samples should be taken first, to avoid contamination from the sample site.
- Travel between local control sites and sampling areas should be minimized because of potential contamination caused by people, equipment, and/or vehicles.

In contrast to a local control site, an *area control site* is in the same area, e.g., a city or county, as the sampling site but not adjacent to it. The factors to be considered in area control site selection are similar to those for local control sites. All possible efforts should be made to make the sites identical except for the presence of the species of interest at the site under investigation [Black]. In general, local control sites are preferable to area control sites because they are physically closer. However, when a suitable local control site cannot be found, then an area control site will still allow important background samples to be collected.

Field Spike Samples

Field spike samples are selected field samples to which a known amount of the analytes of interest are added during their collection in the field [Black]. They are used to identify field, transportation, and matrix effects [Meier]. For complex matrices liable to cause interferences in the analysis, a field spike sample may be used to measure the magnitude of those interferences [Black]. It is important that field spike samples be prepared by experienced personnel so that interpretation of analytical results are not complicated by errors [Dupuy, 1989]. In the event that background samples are not practical to collect, field spiked samples can provide a reasonable substitute that will help to estimate matrix effects; however, field spiked samples may have higher extraction efficiencies than unspiked test samples.

CHAPTER 3

Sampling Water Matrices

The ASTM Committee D-19 lists types of waters as surface waters (rivers, lakes, artificial impoundments, runoff, etc.), groundwaters and springwaters, wastewaters (mine drainage, landfill leachate, industrial effluents, etc.), saline waters, estuarine waters and brines, waters resulting from atmospheric precipitation and condensation (rain, snow, fog, and dew), process waters, potable (drinking) waters, glacial melt waters, steam, water for subsurface injections, and water discharges including waterborne materials and water-formed deposits. Many of these water types require special sampling and handling procedures peculiar to that source.

OBTAINING REPRESENTATIVE SAMPLES

Waters can be very heterogeneous, both spatially and temporally (with time), making it difficult to obtain truly representative samples.

Solids with specific gravities only slightly greater than that of water are usually organic. They will remain suspended in the flow but also will form strata in smoothly flowing channels. Oils and solids lighter than water (usually organic) will float on or near the surface. Some liquids, such as halogenated organic compounds, are heavier than water and these will sink to the bottom.

Variations in Surface Water Sources

The chemical composition of lakes and ponds may vary varies significantly depending on the season. Sometimes the recommended length of the sampling study is about ten times longer than the longest period of interest [Cowgill, 1988]; however, this is often not practical because of sampling costs.

The composition of flowing waters, such as streams, depends on the flow and may also vary with the depth. When a single fixed intake point is used, it should be located at about 60% of the stream depth in an area of maximum turbulence, and the intake velocity should be equal to or greater than the average water velocity [Newburn].

Stratification within bodies of water is common. In lakes shallower than about five meters, wind action usually causes mixing, so neither chemical nor thermal stratification is likely for prolonged periods; however, both may occur in deeper lakes [Cowgill, 1988]. Rapidly flowing shallow rivers usually show no chemical or thermal stratification, but deep rivers can exhibit chemical stratification with or without accompanying thermal stratification. Stratification may also commonly occur where two streams merge, such as the point where an effluent enters a river.

Stratification is also a problem with ocean sampling; various species may be stratified at different depths. In addition, the composition of near shore waters usually differs greatly from waters far from shore. Estuarine sampling is even more complex because stratifications move up rivers unevenly [Cox].

Variations in Precipitation Sources

Precipitation (snow, rain, fog, dew) can vary greatly in composition during the course of a sampling event. Meteorological conditions and variation in the atmospheric concentrations of species of interest cause this. All data involving precipitation samples must include information on meteorological conditions and the sampling time in relation to the total precipitation event [Tanner]. Automated samplers that open when precipitation begins are the preferable collection devices for rain samples [Lodge].

The location of collection devices is especially important when sampling ice and snow. The chemical composition of ice reflects the chemical composition of the surface water and the rate which it forms ice. The dust and/or plankton it entraps has been shown to contribute concentrations of metals such as iron, titanium, and molybdinum. Furthermore, silicon, aluminum, phosphorus, barium, strontium, and manganese (and probably organic contaminants) may show concentration-depth relationships in ice. Therefore, if geochemical (spatially related) data are desired, composite sampling from multiple locations is sufficient, but if data on water composition in relation to the ice in contact with it is desired, the ice must be sampled in a series of strata [Cowgill, 1988].

The chemical composition of snow changes much like that of rain. Initial snow and rain samples usually have higher concentrations of chemical contaminants than samples collected later in the precipitation event. This is because the initial precipitation scrubs dust, other particulates, and water soluble gases (such as H_2S, NO_x, and SO_3) out of the air as it falls. Snow should always be sampled at the time of snowfall because pollutant concentrations usually change significantly on standing [Cowgill, 1988]. Another difficulty with sampling snow quantitatively is that wind can move the snow either into or out of most sampling containers [Lodge]. Fog, clouds, and dew are especially difficult to sample for some of the same reasons.

Variations in Groundwater Sources

Groundwater vulnerability to contamination is affected by water depth, recharge rate, soil composition, topography (slope), as well as other parameters such as the volatility and persistence of the analytes being determined [Dupuy, 1989]. In planning groundwater sampling strategies, knowledge of the physical and chemical characteristics of the aquifer system is necessary (but almost never known). Groundwaters presents special challenges for obtaining representative samples.

Temporal issues need to be considered such as the time of year sampling will be done, whether to sample before or after rainy seasons, etc., and other considerations such as sampling after periods of high agricultural chemical usage. In constructing and using monitoring wells, alteration of the water being sampled must be minimized. Care must be taken during the drilling process not to cross contaminate aquifers with loosened topsoil possibly laden with agricultural/industrial chemicals [Dupuy, 1989]. Well construction and materials can profoundly influence the chemical composition of samples, so material blanks are important.

Purging wells before sample collection eliminates stagnant water. The method and rate of purging, time between purging and sampling, and sampling itself will depend on the diameter, depth, and recharge rate of a well. Each well should be slug, pressure, or pump tested to determine the hydraulic conductivity of the formation and to estimate the extent and rate of purging prior to sampling [Smith, et al., and Barcelona, 1989]. The standard purge volume obtains a stabilized concentration of the parameter of interest [Steele]. Purge volumes usually range from 3 to 10 well volumes. Sometimes changes in pH temperature, or conductance measurements can be monitored in consecutive samples to determine when a sample is representative, i.e., these surrogate values stop changing.

SELECTING SAMPLING DEVICES

The single greatest factor influencing the collection of representative water samples with automatic samplers may be the skill of the user [Newburn].

Sampling devices must be constructed of materials compatible with the matrix and target analytes. Hardware should be stainless steel; plated or painted hardware is not acceptable. Equipment (rinsate) blanks are very important. Usually double or triply-distilled water is used to rinse sampling equipment prior to its use.

Medical grade silicone rubber in peristaltic pumps avoids sample contamination by the organic peroxides used in the manufacturing of conventional grades of silicone rubber [Newburn]. Short lengths of medical grade silicone rubber at the tube compression reportedly does not alter or contaminate samples [Ho]. If organic species are being collected the rest of the tubing should be Teflon®. When sampling for water quality parameters (pH, color, chloride, dissolved oxygen, etc.) polyvinyl chloride (PVC) tubing may be used, but it should be of food-grade quality to prevent phenolic compound contamination of samples [Newburn].

Any sorption of the analytes of interest, in or on the device, must be documented. If such information is not available then analyte sorption with the device must be investigated prior to test sample collection. If the sampling device sorbs the analyte of interest or contributes a significant analytical interference then the samples obviously are not valid, and other means of sampling must be used.

Selection of the sampling device frequently depends on the body of water being sampled. Samples collected from large bodies of water are usually collected manually. Automatic samplers are commonly used for consistent samples of streams and wastewater discharges. Samplers are designed to collect either discrete or composite samples and most are capable of gathering either timed interval samples or samples proportional to flow. Various designs for automatic samplers are available, and selection usually depends on their intended use. Significant selection factors are:

- Intake velocity
- Watertightness
- Electrical or insulation quality
- Explosion proof quality
- Ease of field repair

Glass vacuum pump samplers are usually used to collect samples of dew and fog but the reproducibility is poor. Samples of water analyzed for volatile organics are always grab samples using glass vials with Teflon-lined caps; no headspace is allowed.

Glass containers with Teflon-lined caps should generally be used when organic compounds are the analytes of interest. In contrast, when metal species are the analytes of interest, the samples generally should be collected in plastic (usually polypropylene) or glass containers with added nitric acid for stability [Parr, et al., 1988].

Characteristics of Various Types of Water Samplers

There is no universally accepted sampler, so the selection of sampling equipment must be made to accommodate the goals of the sampling plan [Haile]. Vacuum samplers produce higher biological oxygen demand (BOD), chemical oxygen demand (COD), and solids concentrations than peristaltic pumps. If the strainer of a vacuum sampler is allowed to rest on the bottom of the sampling site, the high intake velocity can scour sediments from around the strainer and enrich the sample [Newburn]. Also, suction lift (vacuum) samplers will cause volatile compounds to outgas and be lost. Another potential problem with vacuum samplers is that their metering chambers can serve as a source of cross contamination between samples due to their relatively large wetted surface areas [Newburn]. However, one advantage of vacuum samplers is that they tend to keep heavy solids in suspension. Another advantage of vacuum samplers with metering chambers, and also peristaltic pumps that can compensate for water level changes, is more accurate sampling when the water level varies significantly from one sample interval to the next.

Discrete samplers can take individual samples, usually at uniform time intervals, and retain them in separate containers for analysis. Two optional modes of operation include nonuniform time intervals and time override of flow-proportioned sampling. Nonuniform time intervals give the option of programming different times between samples. They are useful where variations in flow or analyte concentrations occur [Newburn].

Composite samplers mix samples together in a single container. Their advantage is that many frequent samples can be taken and a time averaged sample is obtained. However, if infrequent events with large concentration variances occur, this information may be averaged out by dilution. A flow-proportioned composite sample, in which small aliquots are collected over small increments of flow, provides the most representative sample of the flow over a given time [Newburn].

Special Considerations for Groundwater Sampling

Select the materials for well construction carefully. Cement used for polyvinyl chloride (PVC) pipe joints can leach into samples from wells; this can be prevented by using threaded pipes. Equipment for monitoring wells should be constructed of stainless steel or polytetrafluoroethylene. These materials cause the least contamination, but they are substantially more expensive than PVC and other more common materials [Kent].

Sampling devices selected for groundwater monitoring should consider the well diameter and yield as well as the limitations in the lift capacity of the devices and the sensitivity of the analytes to construction materials. Groundwater sampling devices should be designed to avoid excessive aeration so that analyte volatilization and oxidation are minimized. Loss or introduction of gases or volatile organics can affect analytes of interest [Kent]. Commonly used devices include electric submersible pumps, bailers, suction-lift pumps, and positive displacement bladder pumps, the latter being generally considered the best for accuracy and precision under many circumstances.

Bailers are often used for both purging and sampling small diameter shallow wells, but they have the disadvantages of mixing, collecting particulates from the well bottom or casing, and aerating or degassing volatile analytes from samples [Kent]. Some of these disadvantages can be minimized by modifying a bailer for a bottom draw valve or a dual check valve and gently lowering it into the water. Another problem is having organics from the air absorbed into the water as it is poured from the bailer to the sample container [Kent]. Thus, field blanks are especially important when using bailers and should always be collected when using this device.

Suction lift and gas displacement pumps often measure the

Table 3. Potential Contaminants from Sampling Devices and Well Casings

Material	Contaminants Prior to Steam Cleaning
Rigid PVC-threaded joints	Chloroform
Rigid PVC-cemented joints	Methyl ethyl ketone, toluene, acetone, methylene chloride, benzene, organic tin compounds, tetrahydrofuran, ethyl acetate, cyclohexanone, vinyl chloride
Flexible or rigid Teflon® tubing	None detectable
Flexible polypropylene tubing	None detectable
Flexible PVC plastics tubing	Phthalate esters and other plasticizers
Soldered pipes	Tin and lead
Stainless steel containers	Chromium, iron, nickel and molybdenum
Glass containers	Boron and silicon

amount of sample delivered inaccurately. In addition, they will cause degassing and the loss of volatile components in the samples [Kent].

COMMON SAMPLING PROBLEMS

Water sample contamination is always a problem, and it increases in importance as the analyte concentration levels decrease. To some extent, contamination sources may depend on the body of water being sampled. For instance, in groundwater monitoring, contamination from well construction materials can be significant and material blanks become very important. However, many potential contamination sources are common to all water samples.

Sampling devices and sample containers are always likely sources of contamination. Carryover between samples from the sampling device must be prevented. Contaminant leaching from sampling devices and containers is very complex and requires serious attention. Table 3 shows the types of contaminants caused by materials used in sampling devices and well construction monitoring [Cowgill, 1988]. Additionally, tin and lead are common contaminants to water transported through soldered pipes. Water containing high calcium levels tends to extract lead preferentially, but tin is removed in small amounts for many years [Cowgill, 1988].

Sampling protocols often recommend that samples that analyze groundwater monitoring wells for metals be field-filtered under pressure before preservation and analysis. Samples collected for metals are usually acidified; acidification of unfiltered samples can lead to dissolution of minerals from suspended clays. Samples to be collected for organic compounds analyses, however, are never filtered [Barcelona, 1989].

Blanks are used to assess contamination. Blank samples usually should include equipment, field, and background blanks; selections should be made by considering all likely sources of contamination for the specific situation.

Analyte sorption is also a common problem. PVC and plastics other than Teflon tend to sorb organics and leach plasticizers and other chemicals used in their manufacture. In addition, some pesticides and halogenated compounds strongly adsorb to glass. When analyzing these substances in water samples, therefore, it is important not to prerinse the glass sample bottle with sample before collection. It is equally important at the laboratory to rinse the sample container with portions of extraction solvent after the water sample has been quantitatively transferred into the extraction apparatus [Dupuy, 1989].

Tubing material used in automatic sampling devices is important; depletion of halocarbons from water depends more on the tube material than on the tubing diameter (surface area). However, when a constant flow rate is used, losses are more likely to occur with an increase in tubing diameter. Thermoplastic materials, e.g., polypropylene, appear to sorb many organic analytes efficiently, so they should be avoided in sampling devices [Cowgill, 1988].

Sorption of metals at low concentrations on container walls depends on the metal species, concentration, pH, contact time, sample and container composition, presence of dissolved organic carbon and complexing agents [Cowgill, 1988]. Preserving metals samples with acid usually prevents this problem.

PVC reportedly containing zinc, iron, antimony, and copper may leach into water samples. Polyethylene has been reported to contain antimony which may leach into water [Cowgill, 1988]. Flexible PVC and plastics other than Teflon usually contain phthalate esters which may also leach into water samples [Kent]. Phthalate esters interfere with instrument sensitivity by masking other contaminants [Coakley].

Variations in the permeability of an aquifer can affect the representativeness of groundwater samples. If the wells have varying recovery rates, varying concentrations of the analytes will result. Vertical gradients of flow between permeable strata within an aquifer can result in samples from multiple zones within one well [Kent].

Finally, groundwater/well water sampling at municipal and domestic wells is best, if possible, at locations prior to any purification/treatment process. This more accurately determines what contaminants are in the aquifer. Chlorination, filters, softeners, and other treatments such as iron, acid, potash, etc., may chemically alter or physically adsorb the analytes of interest. Also, histories and knowledge of any chemical usage in or near wells can provide valuable information. For example, some domestic well owners have been known to pour bleach into their wells as a disinfectant [Dupuy, 1989].

SAMPLE PRESERVATION

The stability of analytes of interest depends on how well the samples are preserved. Preservation instructions must specify proper containers, pH, protection from light, absence of headspace, chemical addition, and temperature control. The chemistry of all analytes must be considered, and it should be recognized that certain reactions, e.g., hydrolysis, may still occur under recommended preservation conditions [Donaldson].

Holding time is the length a sample can be stored, after collection and preservation and before preparation and analysis, without significantly affecting the analytical results. Holding times vary with the analyte, preservation technique, and analytical methodology used. Usually maximum holding times (MHTs) are specified by the EPA, and they must be considered and planned for when sampling and analysis protocols are being developed [Burns].

MHTs of volatile organic compounds are usually 14 days using EPA methods. However, most of these (with the exception of the aromatic compounds that are prone to biological degradation and some highly halogenated compounds that may undergo dehydrohalogenation) have proven stable in water samples for much longer times [Maskarinec, et al.].

Water samples are in a chemically dynamic state, and the moment they are removed from the sample site chemical, biological, and/or physical processes that change their compositions may commence [Parr, 1988]. Analyte concentrations may become altered due to volatilization, sorption, diffusion, precipitation, hydrolysis, oxidation, and photochemical and microbiological effects.

Free chlorine in a sample can react with organic compounds to form chlorinated by-products. Drinking water and treated wastewaters are likely to contain free chlorine. Sodium thiosulfate should be added to remove free chlorine [Parr, 1988]. An exception to this recommendation is when dihaloacetonitriles are to be analyzed; these compounds are degraded by sodium thiosulfate [Trehy and Bieber].

Samples with photosensitive analytes (such as polynuclear aromatic hydrocarbons and bromo- or iodo-compounds) should be collected and stored in amber glass containers to protect them from light [Parr, 1988].

The composition of water samples may also change because of microbiological activity. This is especially prevalent with organic analytes in wastewaters subjected to biological degradation. These samples (and samples containing organic analytes in general) should be immediately cooled, stored, and shipped at low temperature (about 4°C). Sometimes extreme pH conditions (high or low) or the addition of toxic chemicals, e.g., mercuric chloride or pentachlorophenol, are used to kill microorganisms, but this is not common because of their potential for reacting with other analytes [Parr, 1988]. Recent studies have indicated that sodium bisulfite addition may be just as effective for preserving water samples for organic analytes as the addition of hydrochloric acid [Maskarinec, et al.]; however, this is not yet an accepted modification to EPA protocols.

Samples preserved by cooling should be cooled first in a refrigerator or with "wet" ice (frozen water); "blue ice", a synthetic glycol packaged in plastic bags and frozen, is then acceptable for maintaining low temperatures. Initially, blue ice cools less efficiently, and it may take longer to lower sample temperatures [Kent]. A maximum temperature thermometer will document whether temperatures exceeded desired values during storage.

Analytes also may form salts that precipitate. The most common occurrence is precipitation of metal oxides and hydroxides due to metal ions reacting with oxygen. This precipitation is usually prevented by adding nitric acid; the combination of a low pH (less than 2) and nitrate ions keeps most metal ions in solution. Other acids (especially hydrochloric and sulfuric) may cause precipitation of insoluble salts and/or analytical interferences [Parr, 1988].

Waters with cyanides or sulfides require added sodium hydroxide to ensure that hydrogen cyanide or hydrogen sulfide gas is not evolved. Waters with ammonia are preserved by adding sulfuric acid. However, adding of sodium hydroxide or sulfuric acid may precipitate other cations (especially metals), so separate test samples are necessary when cyanides, sulfides, or ammonia are target analytes.

CHAPTER 4

Sampling Air Matrices

Many analytes of interest in air are rather reactive compounds, present in very low concentrations. They may be distributed between two or more phases, e.g., gases and solids or gases and liquids. Proper sampling requires a single phase if inferences are to be made from the chemical composition of samples. However, since most sampling devices immediately perturb the equilibrium of the distribution of analytes between solid and gaseous phases, conclusions about phase distributions from most air samples is tenuous.

Air can contain a large number of organic compounds. As a general rule, organic compounds are present in ambient air in very low concentrations. Sometimes there are higher concentrations inside and around certain industries. Thus, common types of air samples include indoor air, ambient (outdoor) air, and air from stacks or other kinds of emission exhausts (including automotive exhausts).

Another type of air sampling being used more frequently involves soil atmospheres. Air samples from soils over and around landfill waste sites are increasingly being used to assess chemical pollutants in the soil [Steele].

Low concentrations of analytes always complicate the process of obtaining samples for analysis. Analyzing low levels of organic compounds in the air generally requires large sample volumes using solid sorbents for vapor phase compounds and filters for solid (particulate) phase compounds.

Large variations in analyte concentrations over short periods of time are common with air samples. Consequently, unrepresentativeness and variations from sampling may be larger orders of magnitude than those from laboratory analyses. In addition, air exposure concerns include short term acute and chronic risks in addition to the long term exposure risks characteristic of water, soil, and biota matrices. Thus, pollutant action levels at specified concentrations may be in minutes rather than days or years.

OBTAINING REPRESENTATIVE SAMPLES

Vapor pressure and polarity are two of the most important physical properties affecting compound sampling from air. Important contributors to obtaining representative air samples include the efficiency of the collection apparatus, integrity of the sample entering and being removed from the apparatus, location of the sampler, and timing of the collection [Ziman].

For stationary emission sources such as stacks, the sampling site and number of traverse points used for collection affect the quality of the data. Emission tests are based on the assumption that a sample obtained at a given point is representative; if this assumption is wrong it can cause serious problems. Since analyte concentration gradients can be very large, screening analyses are usually needed to measure their values prior to the main sampling effort.

One method of assessing the magnitude of analyte concentration variations due to sampling techniques is to place two or more identical samplers next to one another. Assuming the analytical precision is known, the remaining variation (imprecision) in the data may usually be the result of sampling efforts.

Obtaining representative indoor samples requires special considerations that are different from those involved with stationary sources.

- The ventilating systems must be carefully evaluated. Heating and cooling equipment can alter air flow and add pollutants to the air.
- The location of a sampler within a room will influence the results obtained.
- The usual activities within a room must be considered and documented. A decision has to be made to include or stop these activities during sampling.
- Histories of cleaning activities, pesticide applications, or other activities that could add compounds to the air must be documented and considered.

Each of these considerations is important for minimizing bias and false positives in data from indoor samples [Ferrario, 1989].

If samples are to retain representativeness, gaseous analytes must not react irreversibly with filter or sample container surfaces or with collected aerosol particles. Furthermore, collected aerosols must be handled in ways that will minimize analyte release or retention; however, it is impossible to prevent losses of volatile analytes collected on filters [R.G. Lewis, 1986]. Also, filters must not be contaminated with gaseous materials that can convert to aerosols on the filter media, because this will cause false positives [Ziman].

Sorbent sampling is fraught with contamination, interferences, analyte capture efficiency problems, and recovery efficiency problems. It is also a very important and useful technique when used with appropriate controls. Volatile compound breakthrough before sampling completion will cause errors in their measured concentrations; therefore, two sorbent cartridges are used in series. When no detectable concentration of the analytes of interest (above background) is measured in the second cartridge, no measurable breakthrough has occurred. Also, unknown reactions may occur on the sorbent, during solvent extraction, or during thermal desorption. Using two different sorbents to compare analytical results may provide documentation that unknown reactions did not occur.

Obtaining representative samples from dry deposition (gravitational particle settling and the aerodynamic exchange of trace gas and aerosol particles from air to a surface) requires that the monitoring site be located within an area that is both spatially homogeneous and representative of the larger region of interest. Sampling protocols designed to provide information relative to dry deposition must document meteorological and surface components because, for many chemical species, deposition velocity is as variable a quantity as the concentration.

From the above it is obvious that valid sampling methods, sample stability, appropriate blanks, and control samples are extremely important in verifying that representative samples of air have been taken.

SELECTING SAMPLING DEVICES

Aerosol particles may be sampled by use of appropriate filters; however, various types of impactors for aerosol collection permit size discrete samples. These can be quite useful in transport and deposition studies [Barcelona, 1989].

Solid sorbents are commonly used to collect volatile and semivolatile organic compounds (VOCs and SVOCs). Solid sorbent media may be divided into three categories: organic polymeric sorbents, inorganic sorbents, and carbon sorbents. Both the capture properties and the recovery process must be considered (and verified if they are not documented) when sampling with solid sorbents.

In the vapor phase, nonvolatile organic compounds have negligible concentrations in the atmosphere. They are usually bound to solid particles and can be collected with filtration devices. However, high molecular weight hydrocarbons (above C_{10}) can still be in sufficient concentrations in a gaseous state to be relevant for ozone studies [Ziman].

Surface sampling methods for dry deposition based on micrometeorology provide averages over large spatial scales and typically sample for about 24 hr.

Direct sampling for liquid water in clouds may be done from an aircraft with a multiple slotted rod collector inserted through the skin of the aircraft; the water is collected by gravity inside the aircraft. Similar devices collect water samples from fog.

Devices for Collecting Volatile Compounds

Volatile compounds are generally considered to be below the molecular weight of C_{10} hydrocarbons. They may also consist of oxygenated, sulfur-containing, or nitrogen-containing species. When ambient air samples for these types of compounds are taken, eliminate ozone and nitrogen oxides so that reactions with these oxidants will not take place in the sampling device. Alternatively, experiments in which the sampling conditions are duplicated can demonstrate that the pollutants of interest are inert to those oxidants over the course of sample collection and analysis [Ziman].

Steel canisters, with interior walls specially electropolished to prevent decomposition of the collected organic compounds, are commonly used for collecting ambient air samples. Canisters may be either evacuated in advance and then filled with air or pressurized using pumps with inert interior surfaces.

Tedlar®/Teflon® bags are another type of container used to collect whole air samples. These must always be checked for leaks before use. They are filled using pumps with inert interior surfaces or indirect pumping. With the latter technique, the deflated bag is placed in an airtight container which is then evacuated; the bag fills as it expands in the container.

Samples of soil atmospheres are obtained from boring small diameter holes to a depth of three or more feet. Sampling ports, such as plastic tubes with perforations at their base, are installed in these holes. Air samples are extracted with peristalic pumps and pumped into Tedlar/Teflon bags. Common problems include bore hole preparations and sample handling [Steele].

Condensing volatile organic compounds from air into a cryogenic trap is an attractive alternative to sorbent sampling, particularly when it is combined with ambient air sampling in appropriate containers. The advantages of this technique are that it collects and measures a wide range of organic compounds, greatly reduces contamination problems, and obtains consistent recoveries. Maintain the temperature of the cryogenic trap above the boiling point of oxygen so that the latter is not collected [Lodge].

Passive monitors use a sorbent or reactive medium contained in a protected environment. The principle of diffusion of the analytes, from ambient air into the interior of the monitoring device, is used to calculate their concentrations in the ambient atmosphere from their sampling volumes.

Devices for Collecting Volatile and Semivolatile Compounds

The most widely used procedure for sampling ambient air for volatile and semivolatile compounds is to pass measured volumes of air (typically 2 to 100 L for most VOCs and 2 to 500 m^3 for SVOCs) through a solid material which sorbs the component of interest. Unfortunately, solid sorbents, like most concentration techniques, are not compound specific, and unwanted compounds must be separated from the target compounds with which they are co-collected. In addition, breakthrough of the compounds of interest can occur, resulting in lower than actual measured concentrations. A second solid sorbent cartridge in series placed behind the first will verify that breakthrough did not occur when the analyte concentrations from the second cartridge do not statistically exceed those from the corresponding blanks. However, the sum of the quantities of analytes collected in the two cartridges should not be assumed to represent the total quantity of analytes in the air sample [R.G. Lewis, 1989].

Since most SVOCs are phase distributed, a particle filter is usually placed in front of the sorbent cartridge. However, the distribution of SVOCs between the filter and the sorbent will not accurately reflect atmospheric phase distributions, so the filter and sorbent are combined for analysis [R.G. Lewis, 1986].

After collecting the compounds of interest, solvent extraction or thermal desorption is used to recover organic compounds from the sorbent. Contaminants from the sorbents in these devices commonly interfere with subsequent analyses, thus they must be thoroughly precleaned before use, and blank samples must be analyzed to document background levels of potential interferants.

The advantage of solvent extraction is that only an aliquot of the extract is analyzed. This makes replicates possible, and the concentration of analytes can be optimized. The principal disadvantage of solvent extraction is that only small aliquots (typically 0.1 to 1%) of the extract can be analyzed, resulting in higher levels of detection and quantitation than could be achieved if all the sample were used. Collecting large volumes of sample may counteract this disadvantage [R.G. Lewis, 1989].

The advantage of thermal desorption is that the entire sample is used in the analysis, so smaller air volumes suffice. This can be a disadvantage if replicates are needed [Clements, 1988]. Another disadvantage is that the analytes may be decomposed by pyrolysis [Lodge]. Also, thermal desorption is not possible or practical with most semivolatile compounds [R.G. Lewis, 1989].

Sorbent Sampling Materials

Inorganic sorbents include silica gel, alumina, magnesium aluminum silicate (Florisil), and molecular sieves. These sorbents are more polar than the organic polymeric sorbents. They more efficiently collect polar organic compounds; unfortunately, water is also efficiently captured, and this causes rapid sorbent deactivation [Clements, 1988]. In addition, isomerization of organic compounds may be catalyzed by some of these sorbents [Lodge]. Therefore, inorganic sorbents are not often used for sampling volatile organic compounds.

Activated carbon sorbents are relatively nonpolar compared to the inorganic sorbents, and water is less of a problem. Although, water may still prevent analysis in some cases, especially where the relative humidity is high. The irreversible sorption of some organics to activated carbon and the potential reactions promoted by the high surface area of this sorbent also may cause problems [Clements, 1988].

Organic polymeric sorbents include materials such as a porous polymeric resin of 2,4-diphenyl-p-phenylene oxide (Tenax®-GC), styrenedivinylbenzene copolymer (XAD) resins, and polyurethane foam (PUF). These materials have the important feature of collecting minimal amounts of water in the sampling process.

Tenax sorbents are widely used in sampling ambient air for

organic compounds (Tenax-GC, a polymeric solid sorbent originally developed as a support for gas chromatographic columns, is the most widely used solid sorbent). Tenax-GC has a low affinity for water and a high thermal stability, which permits the thermal desorption of collected volatile materials [Clements, 1988].

Polyurethane foam is often used for sampling semivolatile organic compounds such as pesticides and polynuclear aromatic hydrocarbons. One advantage is its low resistance to air flow; large air volumes can be passed through it. It also provides good desorption recoveries with common solvents.

Devices for Collecting Nonvolatile Compounds

Nonvolatile compounds, i.e., elemental carbon and compounds with high molecular weights, have negligible concentrations in the vapor phase; these compounds are often bound to solid particles. They are sampled in the usual methods for atmospheric particles, e.g., filtration. High-volume samplers are commonly used as well [Clements, 1988].

Dry deposition can sometimes be measured by intensive micrometeorological techniques, but these methods require either demanding chemical precision for the gradient method or rapid frequency response for eddy correlation. For routine measurements, less direct approaches are frequently used, ones in which dry deposition is computed from atmospheric concentration and site-specific deposition velocity data. Available information limits such simplified approaches to only a few gaseous chemical species, e.g., ozone, sulfur dioxide, nitric acid, particulate sulfate, nitrate, and ammonium; usually average concentration data can be used [Hicks, et al.].

Filter packs are sometimes used as prefilters for sampling air. Teflon or similar prefilters that remove particles from the airstream may be followed by a nylon filter to remove nitric acid. A cellulose final filter treated with potassium or sodium carbonate or bicarbonate (and sometimes also with glycerol) can ensure a moist surface [Hicks, et al.].

Impingers (bubblers) commonly have a liquid collection medium in which chemical speciation, e.g., the valence form of a metal, can be changed substantially. Impingers are not recommended for long term, routine monitoring because evaporation of the collection fluid can be significant [Hicks, et al.].

COMMON SAMPLING PROBLEMS

Loss of analytes or reduced concentrations from irreversible sorption on the walls of sampling containers can cause very serious problems. Sorption or penetration of analytes with Tedlar bags may occur, especially at the typical, very low ambient air concentrations [R.G. Lewis. 1989].

Cloud droplets, aerosols, and gaseous phases are usually separated on the basis of differences in aerodynamic properties. An aerosol is a colloidal suspension of particles in a gas [Lodge], and these particles should be sampled as near to isokinetic conditions as possible, i.e, with no change in air speed or direction, if the original aerosol size distribution is to be maintained. Isokinetic sampling is especially critical if coarse particle or cloud droplet analyses are to be performed. However, this is not usually possible in ambient air since most samplers are designed to collect a representative sample at a velocity less than 20 km/hr [R.G. Lewis, 1989]. In the presence of a cloud/liquid water phase, most water soluble aerosol particles are incorporated into the cloud water phase.

All micrometeorological methods involve an assumption that the fluxes measured at the height of the turbulence sensors are the same as the fluxes to the underlying surfaces. To assure the validity of this assumption, measurements must be conducted only in conditions that do not change with time, and at locations where the surface is horizontally uniform [Hicks, et al.].

INFLUENCE OF METEOROLOGY AND TOPOGRAPHY ON SAMPLING AIR

The primary meteorological effects that must be considered when obtaining air samples are: wind direction, wind speed, temperature, atmospheric stability, atmospheric pressure, and precipitation.

Wind direction is the most important factor, and it must be constantly documented during air sampling activities. A change in wind direction can provide large variations in the amounts of air pollutants collected in just a short period of time.

Wind speed may affect the volatilization rate of contaminants from liquid sources and also the concentrations of pollutants downwind. As the wind speed increases, it may increase the volatilization rate of some liquids; however, increased wind speed also usually dilutes the concentration of vaporized pollutants downwind of a source. On the other hand, greater wind speed may increase the concentration of nonvolatile contaminants sorbed to particulates such as soil and dust. This is because, under windy conditions, larger particulates and greater quantities of them can be transported from contaminated sites.

Another factor that increases volatilization is higher temperature. Both solar radiation and air temperatures must be considered, but in general, solar radiation usually has the larger influence causing the volatilization of liquids with high vapor pressures.

Atmospheric instability relates to vertical motions of the air. In unstable atmospheric conditions, dispersion of contaminants in the air increases throughout various vertical levels. Downwind contaminant concentrations are usually higher when stable atmospheric conditions exist.

Atmospheric pressure affects the migration of gases through and out of landfills. Volatile contaminants from landfills may be released at higher rates during periods of low atmospheric pressure changes. The reverse may also occur, so slower volatilization rates may be observed when high atmospheric pressures are in effect. However, if significant lag times are associated with these pressure changes they may not be very noticeable, and each landfill must be considered separately.

Precipitation decreases overall airborne contaminants, although contaminant concentrations will usually be highest in the initial precipitation samples. During precipitation events, contaminants are removed from the air when particulate matter is physically carried down with the precipitate and gaseous contaminants are dissolved in it. Particulate matter transported by the wind is also usually insignificant when the soil or other particulates are wet. Also, volatile contaminants emitted from wet soils may be at much lower rates than from dry soils; however, the humidity of the air at less than 100% generally has little effect on either volatilization or transport of contaminants in air.

TOPOGRAPHY EFFECTS

Mountains, hills, valleys, lakes, and seas can significantly affect the wind direction and the amount of mixing or dispersion of contaminants in the air.

Within valleys, the air generally flows along the axis of the valley with a bi-directional distribution. During evenings and nights, colder air from high slopes may drain down into a valley causing a thermal inversion of the air. This releases a concentration of contaminants from valley sources because of decreased vertical dispersion. If the valley itself has a slope, the colder air will tend to flow down that slope and carry the contaminants with it. The reverse situation can also occur in the morning and daylight hours, but this phenomenon is not as frequent; the up-valley and up-slope winds are not as strong as the cooler down-valley and down-slope winds.

Near the sea coast and large lakes, temperature differences resulting from daylight and nighttime effects also cause bi-directional wind changes. Generally, during the day, the wind will move from the cooler body of water toward the land and then exhibit an upward motion over the land. The reverse situation usually occurs at night when rapid cooling from the land causes air over the land to become cooler than air over the water. These "land breezes" usually are not as strong as the sea or lake breezes.

CHAPTER 5

Sampling Biological Matrices

Sampling biota for chemical analysis presents unique challenges because of the vast size differences between species, variations within a study population, species mobility, and tissue differentiation. When sampling biota, factors to be considered include the purpose of the study, homogeneity of the matrix, concentration of the analytes, efficiency of extracting and concentrating, and the sensitivity of the method to be employed. The nature of the organism, size of the population under consideration, availability, and cost of the material to be studied are factors that often complicate biota sampling. (Sampling human tissue involves similar principles, but the techniques are often very specialized and beyond the scope of this publication.) In addition, the physical characteristics of the analyte, such as phase, volatility, oxidation state, and chemical properties, have pronounced influences on sampling. These characteristics help determine when, where, and how much material must be removed from the environment to ensure sample validity.

Additional factors to be considered when sampling biological materials, especially when animals are target organisms, include [Ferrario]:

- The mixed function oxidase activity in a target organism and how it may affect the analytes of interest
- Bioaccumulation of the analytes by the target organisms
- Whether the analytes of interest exist in the target organism as the parent compounds or as metabolites

- Whether the target organisms are migratory
- Whether there are seasonal, feeding, spawning, or other periodic activities that may influence concentration or location of the analytes within an organism

When referring to the homogeneity of the matrix to be sampled, the matrix itself and also the distribution of the analyte throughout that matrix must be considered. If application uniformity is the subject of interest, each sample must be analyzed individually, and the sample size will approach that of a subsample. If the mean application is the subject of interest, e.g., as in yield studies, then all the individual samples can be composited, the composite can be homogenized, and a subsample can be taken for analysis.

The following considerations are important when dealing with biomaterials:

- Since all or perhaps a large part of the population being studied may eventually be taken, a first consideration is whether the site should be marked or photographed for future reference before being disturbed. For many monitoring or field research studies, a map indicating sampling sites or coordinate designation is sufficient.
- Decisions must be made as to whether the sample will have to be examined macroscopically or microscopically for deposits, wounds, burns, lesions, or pathology. Many of these conditions are indicative of chemical contamination, or conversely, are symptomatic of diseases similar in appearance but unrelated to chemical exposure.
- If any nonhomogeneous reductions need to be performed, that an untimely or inappropriate procedure would complicate, these must be fully documented. Frequently, plant parts are separated in the field, but at a point distant from the actual collection site. Homogeneous reductions may be similarly facilitated when large units, e.g., melons or squash, are held temporarily at ambient temperature and then sectioned before portions are bagged and preserved.

OBTAINING REPRESENTATIVE SAMPLES

Large variabilities and corresponding uncertainties often occur when sampling biological matrices. In many cases, statistical manipulation alone cannot reduce variability to manageable proportions; therefore, nonstatistical strategies often must be used. The more that is known about the target population before sampling, the more readily the effects of variability can be taken into account [Alabert and Horwitz].

Each individual in a biological population should have an equal probability of being included in the sample. The goal of sampling is to reduce this population to manageable and representative portions, aliquots, or pieces. Homogeneity of the selected samples is crucial for minimizing overall variability. The sample planner must decide what should be measured and how accurately. The sample planner also must decide which pieces or parts of the biota should be taken for samples, and equally important, what portions of a biological specimen should not be included in the samples. Documentation of the reasoning behind such decisions is extremely important [Alabert and Horwitz].

Individual biological specimens included in a sample should be chosen at random from a well defined population. Homogenization and common sense can help decrease the variability contribution from within sample unit variability. In addition, statistical analyses of the data may measure variability within samples.

The distribution of values of analyte concentrations in a biological sample population is often Gaussian (bell shaped), so estimates of the average and standard deviation will characterize most data. Besides being useful, the assumption of a Gaussian distribution is often realistic; even if a population is not Gaussian, its linear averages tend to become Gaussian-like [Alabert]. The larger the sample size, the more Gaussian the population of averages becomes (except for the extremes). That is unless the initial population shows heteroscedasticity and auto-correlation in space/time of extreme values [Journel, 1989].

SELECTING SAMPLING DEVICES

Sampling devices will influence *a priori* what types of organism samples will be caught; not every species will have the same probability of being caught by the employed sampling technique [Ferrario].

Among various techniques for collecting fish, electric shockers and slat boxes usually present advantages over hoop nets, gill nets, or trot lines. The latter devices obtain fewer samples per unit time, and they may even kill the specimen before retrieval. Additional sampling devices include seines, dip nets, trap nets, trawls, and pound nets. Traps, dart guns, and nets capture animals for sampling. Botanical sampling tools include saws and pruning shears.

COMMON SAMPLING PROBLEMS

Sample Size

For biological materials, the limiting factor may often be the availability of substrate. The analyte concentration may have a significant bearing on the sample size. If the analyte is present in very low concentration, the sample needs to be large enough to provide a measurable quantity of analyte. Whatever the composition, the primary sample must be of sufficient size to be representative of the system or population (unless targeted sampling is performed). Generally primary samples are too large for individual chemical or physical analyses. These samples must be reduced to replicates of a test size that can be analyzed in the laboratory. The manner in which this reduction takes place is critical to the eventual validity of the data.

In size reductions of nonhomogeneous samples, a selected portion is isolated for analysis because it is either of special interest, represents the known accumulation site, contains the regulated commodity, or represents a limitation of the method. It can be argued that this separation process is merely generating more primary samples and does not really constitute a reduction. In instances where this takes place directly on site, as in the picking of fruit or foliage or the removal of blood or tissue specimens from captured and released animals, the argument is probably valid. At other times, segregating the primary sample into differentiated tissues is an intermediate step performed in the laboratory or at selected facilities.

In size reductions of partially homogeneous samples, the sample (primary and secondary) can be homogenized by grinding, milling, blending, chopping, mixing, or similar physical processes. Further subsampling by dividing, riffling, or taking aliquots will produce uniform portions of a size that is amenable to the analytical method and representative of the whole. The size of the final subsample will be determined primarily by the concentration of the analyte, sensitivity of the method, and capability of the analytical equipment.

Sample Preservation

Preservation should be kept as simple as possible. Freezing is a logical first choice for retarding the decomposition of most analytes. Usually biological samples are placed and held in closed glass containers with inert seals or cap liners [Spittler and Bourke]. Spiking standards into uncontaminated control samples determines preservation effectiveness. The control samples are held with the field samples for later analysis to measure analyte disappearance.

If freezing is the method chosen, whole samples should be frozen if possible. If aluminum foil is used to wrap the samples, the shiny side should not contact the sample because it is coated with a slip agent. Also, some aluminum foils contain trace levels of mercury, so they should not be used if mercury is a target analyte. The foil should be washed with a solvent (such as acetone) first, and then thoroughly dried [Norris].

Many older schemes for organic pesticide analyses recommend that samples be extracted with organic solvents as soon as possible, and that the organic solutions be stored, pending chromatographic or spectrophotometric analysis. Before this option is considered, however, the analyte storage stability in the selected solvent should be ascertained.

CHAPTER 6

Sampling Solids, Liquids, and Sludges

Previously collected monitoring data are often useful guides for selecting and analyzing additional samples. Limited sampling to confirm the validity of existing data may be a useful and cost effective first step in assessing soil or waste contamination. Preliminary liquid screening analyses are less useful, although they may help identify analytes of special concern and assist in selecting sampling and analytical methods.

The format for presenting the monitoring data, e.g., points on maps, contours, analyte levels and/or variations in confidence limits due to sampling density, charts for the overall program and/or subsets of the program, and aggregated data sets, should be determined first when designing the sampling plan. If maps are to be used, the availability of adequate maps with sufficient resolution and validated accuracies must be assured. Using professional surveyors may sometimes be necessary to identify sample locations with high accuracy [Haile].

The frequency of sampling and the number of individual or composite samples needed depend on the time or spatial variability of the individual sampling sites and the precision required. A rigorous geostatistical approach should control the degree of alteration (usually smoothing) of the mapping and contouring interpolation step. A "conditional simulation" technique can avoid such smoothing [Journel and Alabert].

OBTAINING REPRESENTATIVE SAMPLES

Because sample heterogeneity often causes problems in soil, solid, sludge, and some liquid matrices, representativeness uncertainties frequently far exceed the inherent collection and analysis uncertainties. Often it is not possible to quantify the analyte concentration uncertainties associated with the sample selection. In these instances, qualitative descriptions of the uncertainties due to sampling limitations should be clearly described and the associated assumptions fully documented.

Sometimes samples of solids, liquids, and sludges are deliberately collected unrepresentatively. Initial studies at a waste site may focus on the most obviously contaminated areas. Although such samples will not represent the average condition, they will establish the worst case concentrations of the analytes of interest. However, even in these situations it is important to obtain background samples of the environmental matrix (water, soil, air) from either local or area control sites.

Variability arises from the heterogeneity of samples, the size and distribution of the sampling population, and the bias of the sampling and analysis methods. Statistically acceptable values may deviate from the true values because statistically obtained samples include irrevelant data, use inappropriate sampling and analysis methods, or systematically exclude parts of the population [Triegel].

Because many samples are heterogeneous, it is best to select as large a test sample as practical for preparation. An extract or digested solution will be more homogeneous, and it will provide more reproducible aliquots than a smaller portion of the sample. Thus, to extract a sample, a tared bottle with the solvent or acid may be prepared so the sample can be added to it. This immediately starts sample extraction while preserving it [Bone].

Composite samples may help overcome the lack of homogeneity over time or in the distribution of chemical species. At the same time, compositing may dilute peak values of concern. Therefore, if peak concentrations of analytes are important, compositing should be supplemented with grab samples taken at sites and times where high values are suspected.

Liquid waste samples in shallow containers or pools on the ground may be poured or scooped into a sampling container. When liquids in deep containers, such as barrels, are sampled, it is usually best done with a long hollow tube. When the tube is inserted into the barrel and the top of it is closed off, a column of the liquid can be withdrawn and placed in the sampling container. The advantage of this technique is that it provides a fairly representative sample of the liquid at all depths. This is especially important if the barrel contains stratified layers or multiple immiscible liquids.

Whenever possible, statistical analyses should guide sample set collection and analysis; given the problems of heterogeneity, often this is not possible. If the matrix can be stratified into homogeneous components, random statistical sampling within strata can be considered. However, confirmation of component homogeneity frequently may be so difficult as to obviate the benefits of random sampling.

Replicate samples for complex matrices are best prepared in the laboratory rather than the field. In the case of solid wastes, homogenization procedures that ensure the similarity of duplicate samples can inadvertently degrade the sample via induction of chemical reactions or dilution of contaminant levels.

Spiking field samples (adding known amounts of analytes) is often not possible. Considerable effort must be devoted to preparing synthetic QA samples that will approximate the characteristics of the field samples. Spikes from field or synthetic samples, not exposed to the same weathering or other external processes as the analytes of interest, may have higher recoveries than the analytes in the field samples [Piwoni]. Blank samples for soil analysis should be solid whenever possible. They should be placed in the sampling device, then recovered and handled as the samples [Steele].

SELECTING SAMPLING DEVICES

Sediment Sampling Devices

Sediment sampling usually involves devices that drill cores or dredge bottom grab samples. Devices that obtain bottom grab samples are easy to use and can obtain large volumes. Their disadvantage lies in the loss of fine particulates carried away by outflowing water. They collect only disturbed samples and should be used only after the sampling of the overlying water has been completed.

Hand core samplers retain fine particles, their main disadvantage is the small area of the bottom covered; core samples usually require larger numbers than bottom dredge samples. The advantage of core samplers is that they maintain the vertical integrity of layers underlying the sediment—maintaining vertical integrity of easily resuspended sediment layers is usually not practical. Also, most can be adapted to hold inert liners.

Gravity corers have replaceable tapered nosepieces on the bottom and balls or other types of check valves on the top, so water passes through them on descent but doesn't wash out during recovery. These collect mostly undisturbed samples with all but surface strata left intact. They also can be adapted to hold inert liners.

Scoops and trowels are simple and quick but usually collect only disturbed samples. These are well suited for compositing a series of grab samples.

Soil Sampling Devices

Soil sampling devices should be chosen after considering the depth, the soil characteristics, and the nature of the analyte of interest. Surface sampling may be chosen for recent spills or contamination and small migration rates of the analyte. If the analyte of interest is volatile or has been in contact with the soil for a long period of time, sampling at greater depths may be necessary. Soil characteristics will determine the migration patterns of the analytes of interest and the characteristics of the usable sampling devices. The nature of the analyte being sampled, e.g., whether it is volatile or soluble, will influence the sampling depth, the sampling device, and sometimes the materials from which the sampling device must be constructed.

When sampling soil at its surface or at shallow depths (less than about 15–30 cm) scoops or shovels may be used; however, they do not obtain very similar samples. Also these tools are not suitable for sampling soil contaminated with volatile materials, since they may volatilize during sampling and make the samples unrepresentative. As with all sampling devices, careful attention to construction materials is necessary. Generally, scoops and trowels should be stainless steel for soils contaminated with organics and high density polyethylene for soils contaminated with inorganic species.

Sampling devices must be decontaminated between successive samples to avoid cross contamination (the decontamination produces QC samples called "equipment blanks"). Sometimes, when using scoops or trowels, it may be easier to use separate devices for each sample and then have them decontaminated in a lab or other facility equipped for that purpose.

A soil punch or other thin-walled steel tube device is more suited for obtaining reproducible samples at the soil surface or shallow depths. These devices are pushed into the soil to a desired depth and retain a sample. The sample may be removed for compositing or transferred to another sample container. Some thin-walled tube samplers are designed as a combination sampling and shipping device, since the ends of the sampler can be sealed for shipment after the outside of the device is decontaminated.

Sampling at depths greater than one foot requires different techniques and devices. Trenching can obtain analyte profiles, however, it usually costs more than other techniques available. Trenches should be excavated approximately one foot deeper than the desired sampling depth. A soil punch or trowel can dig laterally into the exposed soil to obtain the samples.

Augers, both powered and nonpowered, are also useful in obtaining solid samples from depths greater than about one foot. Augers come in different sizes, and samples may be obtained directly from the auger cuttings. However, this technique can introduce cross contamination between soil layers, contamination from drilling material, non-reproducibility in sample size, and loss of volatile components. A more desirable technique is to reach the desired sampling depth with an auger, and then obtain the sample with a soil probe or split barrel sampler. Soil cuttings should be carefully removed after drilling to avoid cross contamination between soil layers.

Soil probes and split barrel samplers work in a similar fashion. The device is driven into the soil to the desired depth and retains the sample as it is withdrawn. A soil sample obtained in this manner may then be transferred to a separate container for shipment to the laboratory. Stainless steel or Teflon liners are available for split barrel samplers to minimize adsorption of or reaction with analytes. Some of these devices are designed to be sealed for shipment to the laboratory after the exterior is decontaminated.

COMMON SAMPLING PROBLEMS

Sampling that disturbs the chemical composition of some wastes can be dangerous, and appropriate safety precautions as outlined in Chapter 1 are essential. Some wastes may even be explosive when disturbed.

Volatile compounds lost during the sampling and handling of solid and semisolid materials is of special concern. In some cases, samples acquired exclusively for volatile analysis, using special sampling and handling procedures, may be appropriate.

Once sampling teams are in the field, collecting additional samples may be relatively inexpensive, and supplementary collections should be considered in case there are changes in the analytical scheme later.

Rigorous preservation of undisturbed solid and semisolid samples for future analysis should be assured for additional collections.

SAMPLE PRESERVATION

In general, sample containers should be tightly sealed as soon as the samples are taken; headspace should be minimized, and the samples refrigerated as soon as possible. The refrigeration should be maintained until analysis, and the samples analyzed as soon as possible [Bone].

If extraction or acid digestion is required, these procedures should be carried out as soon as possible. That way the extracts or digested solutions can be held for the prescribed holding times.

Either entire core samples or large portions of them should be shipped to a lab, wrapped in solvent washed and dried aluminum foil, or sealed in glass bottles. One pint, wide mouth bottles are useful for core samples—the samples can be cut so that they nearly fill the bottles [Bone].

The most frequent changes in solid, liquid, and sludge samples are volatile loss, biodegradation, oxidation, and reduction. Low temperatures reduce biodegradation and sometimes volatile loss, but freezing water-containing samples can cause degassing, fracture the sample, or cause a slightly immiscible phase to separate. Anaerobic samples must not be exposed to air [Bone].

There are many variations to obtaining good and representative samples of these and the previously discussed matrices. Sampling site conditions often may require minor or major modifications to sampling plans. The key to success lies in keeping the goals of the project in perspective and communicating frequently with all parties involved. Success also requires a solid background of experience and knowledge, a sound QA/QC program, and last but not least, common sense. Deficiencies in any of these attributes makes the job a whole lot harder. A good knowledge of the principles of environmental sampling should help to eliminate some of the many problems inherent in this complex task—the rest remains in the hands of the sampler.

II

ENVIRONMENTAL ANALYSIS

INTRODUCTION TO ENVIRONMENTAL ANALYSIS

Reliable analytical measurements of environmental samples are the next essential ingredient, after sampling, in making sound decisions about safeguarding public health and improving the quality of our environment. To produce high quality data for environmental analytical objectives you must know what analytes need to be measured and what has to be done to obtain good and reliable data, i.e., you must understand the measurement process.

This section provides principles and practical guides for many diverse applications. It is intended to aid in the evaluation of the analytical measurements of environmental samples so that intelligent choices can be made. The many options available in designing and conducting those measurements range from semi-quantitative screening analyses to strict quality assurance programs intended to document the accuracy of data for regulatory enforcement or legal purposes. In this respect, these principles are a set of practical guidelines for making decisions—for planning and executing analytical work, or conversely, for evaluating the usefulness of previously generated analytical measurements.

The principles and guidelines discussed here are organized according to a general operational model for conducting analytical measurements of environmental samples. This section identifies the factors required to obtain reliable data as well as the factors that must be avoided. These factors become important in determining analytical protocols.

Many options for sample preparation and methods of analysis are available for environmental samples. Differing degrees of reliability, as dictated by the objectives, time, sample matrix, and resources available will influence the selection of the protocols chosen.

Analytical protocols for environmental samples frequently require reliable measurements at very low levels, this is significantly different from the analytical protocols for many other types of samples. Often, specific analytes need to be measured at parts per billion or lower levels in complex matrices. Advances in analytical methodology continue to lower the levels at which reliable measurements can be made, while at the same time, requirements for lower measurement levels provide an incentive for future analytical advances. At these extremely low levels, many factors that are of little or no concern in other analytical measurements are of critical importance in influencing the outcome and reliability of environmental analyses.

CHAPTER 7

Planning Analytical Protocols

Careful planning is as essential a part of environmental analysis, as it is with environmental sampling. Inadequate planning can lead to biased, meaningless, or otherwise unreliable results; good planning, on the other hand, will usually produce valid and useful results. The intended use of data measurements should be addressed explicitly in the sample planning process and reiterated in the analytical planning process. Valid and useful results are those that answer a question or provide a basis on which a decision (for example, to adjust a process or to take regulatory action) can be made. The objective of planning is to define the problem and analytical program well enough so that the intended results can be achieved efficiently and reliably.

Never assume that the person requesting an analysis will be able to define the objectives of the analysis properly. It is not unusual for the end user of the analytical data to be unfamiliar with the details and problems of environmental analyses. Numerous discussions between the analyst and the one who requests the analysis may be necessary to agree on what is required, how the results will be used, and what the expected results may be. The analytical methodology must meet realistic expectations regarding sensitivity, accuracy, reliability, precision, interferences, matrix effects, limitations, cost, and the time required for the analysis [Keith, et. al, 1983].

One of the most important elements in planning is the incorporation of essential decision criteria into the overall analytical protocol.

This protocol, which describes the analytical process in detail, should include the overall goals and the data quality objectives, a description of the quality assurance and quality control requirements, the sampling plan, analytical methods, calculations, and documentation and report requirements [Keith, et al., 1983].

Selection of the optimum analytical method is one of the most important factors influencing the reliability of data [Keith, 1980]. In addition to the obvious limitations of equipment availability, sample amount, time, and resources, other factors significantly affect the cost and reliability of the data. There are essentially five data quality objective decisions:

1. The level of confidence required to identify the analyte must be chosen. For example, depending on the consequences of an incorrect decision, a high or low probability of correct identification may be required. Confirmatory analysis with an independent measurement technique may achieve high levels of confidence at higher expense. A lower confidence level can be achieved at less expense by comparing the spectral, chromatographic, or other physical/chemical properties of the analyte with values reported in the literature. Computerized spectral matches also can achieve lower confidence levels at less expense.
2. Deciding on the expected analyte concentrations often affects the method of selection, the amount of sample to be taken, the degree and type of sample pretreatment (e.g., cleanup and concentration), and the method of analysis. Generally, the lower the levels of concentration measurement are required, the higher the cost of doing the analyses.
3. A high level of quality control, the use of validated methods, reference laboratories, and experienced personnel, still does not assure the production of reliable analytical results. All analytical work must be monitored by a system of quality control checks to verify that the results have a given probability of being correct. Ruggedly tested methods with a history of low outlier production, performed by experienced personnel, will usually require a lesser degree of quality control than complex new procedures and inexperienced personnel. However, the most important factors determining the necessary level of quality control are the consequences of being wrong.
4. The degree of confidence regarding analyte concentration will influence the method selection and the number of samples taken, as well as the design of the quality control program that controls precision or accuracy of the data. In general, the higher the degree of precision or accuracy needed, the more rigorous the quality control program must be and the higher the analysis will cost. Eventually the quality control effort reaches a point of rapid increase as lower analyte concentration is approached and further quality control effort produces diminishing returns.
5. The necessary method validation must be chosen. On the basis of the specifications developed in the first three items, the method must be examined to determine whether it can produce the degree of specificity, preci-

sion, and accuracy required. If it does not, the method must be improved or another must be chosen. The first stage of method validation usually involves only intralaboratory validation. Interlaboratory validation usually provides a more realistic representation of specificity, precision, and accuracy. However, intralaboratory verification provides very useful information on single laboratory performance.

If an analytical result is to be used in a screening program or to adjust a process parameter, an unvalidated analytical method may be sufficient and appropriate. On the other hand, if regulatory compliance is the reason for an analysis, a validated analytical method approved by the regulator (especially those referred to in the Code of Federal Regulations), is usually required. In addition to making cost effective decisions, a separate set of decisions must be made within a technical framework [Keith, 1980]. These involve sample preparation and analysis.

A new, efficient approach for selecting the best method for specific analytical needs uses an easy "free text" computer searching program with summaries of methods by analyte in an electronic book. EPA's *Sampling and Analysis Methods Database,* has 150 U.S. EPA methods with hundreds of organic and inorganic analytes that can be searched by any combination of words for analyte, matrix, instrumentation, interferences, etc. (Keith, 1990).

CHAPTER 8

Quality Assurance and Quality Control

Quality assurance (QA) has been described as a system of activities that assures the producer or user of a product or a service that defined standards of quality with a stated level of confidence are met. *Quality control* (QC) differs in that it is an overall system of activities that controls the quality of a product or service so that it meets the needs of users [Taylor, 1987]. In others words, QC consists of the internal (technical), day to day activities, such as use of QC check samples, spikes, etc., to control and assess the quality of the measurements, while QA is the management system that ensures an effective QC system is in place and working as intended.

A laboratory QA/QC program is an essential part of a sound management system. It should be used to prevent, detect, and correct problems in the measurement process and/or demonstrate attainment of statistical control through quality control samples. The objective of QA/QC programs is to control analytical measurement errors at levels acceptable to the data user and to assure that the analytical results have a high probability of acceptable quality.

The data quality is ordinarily evaluated on the basis of its uncertainty when compared with end-use requirements. If the data are consistent and the uncertainty is adequate for the intended use, the data are considered to be of adequate quality. When analytical results are excessively variable or the level of uncertainty exceeds the needs, the data may be of low or inadequate quality. The evaluation of data quality is thus a relative determination. What is high quality in one situation could be unacceptable in another [Taylor, 1987].

QUALITY ASSURANCE

Several basic concepts are involved in quality assurance: (1)planning to define acceptable error rates (the DQO process), (2) quality control to establish error rates at acceptable levels, (3) quality assessment to verify that the analytical process is operating within acceptable limits, (4) reporting and auditing data quality, and improved user feedback to communicate better with the laboratory.

Each laboratory should have an independent quality assurance officer who reports to top management and carries out the laboratory QA program. Furthermore, this QA program should be documented in writing as a QA manual or a QA program plan (QAPP) [Dupuy, 1990].

In addition, every monitoring program should contain an appropriate quality assurance plan and require all participants to follow it strictly. In each case, all involved personnel, including sampling and analytical personnel, the QC officer, and the required statisticians, should develop the QAPP as a joint effort.

The elements of an acceptable QA program include:

- Development of and strict adherence to principles of good laboratory practice (GLP)
- Consistent use of standard operating procedures (SOPs)
- Establishment of and adherence to carefully designed protocols for specific measurement programs

QUALITY CONTROL

The consistent use of qualified personnel, reliable and well-maintained equipment, appropriate calibrations and standards, and the overview of all operations by management and senior personnel are essential components of a sound quality control system. When properly conceived and executed, a quality control program will operate a measurement system in a state of statistical control; random error rates are kept at acceptable levels and have been characterized statistically.

Statistical control is the first requirement that must be met before accuracy can be assessed. The term statistical control is somewhat of a misnomer. It does not mean that statistics controls the process, rather there is statistical evidence that it is in control. Statistical control results from quality control of a measurement process. All critical variables are controlled to the extent necessary and possible; the process is stabilized and the data is reproducible within defined limits. While essentially mandatory, statistical control may be difficult to prove. Continuously monitoring a system to demonstrate its stability is the best, if not the only, approach. Evidence of attainment is best based on the results remaining within the statistical bounds defined by needs of the the data user. Appropriate control charts are the best way to document statistical control [Taylor, 1987]. Control charts should be maintained and used, to the extent possible, in a real time mode. The strategy of the decision process, the corrective actions to be taken when lack of control is indicated, should be planned and followed.

Statistical control is necessary to evaluate the precision of a process. It says nothing about bias, but it is a prerequisite for the evaluation of bias. It is useless to try to identify bias until statistical control has been attained and the process variability has been evaluated. Statistical control does not indicate that a measurement process has been optimized, only that it has been stabilized. For example, similar measurement processes may be operating in a state of statistical control in several laboratories, but the respective performance parameters may differ significantly. The individual precisions may differ and application related biases may be present; however, their mere existence cannot be demonstrated unless each laboratory has attained statistical control. Only then is it possible to seek corrective actions that could enhance greater compatibility of the various data outputs [Taylor, 1987].

Quality control samples are the primary means of estimating intralaboratory variability. The current EPA laboratory QC requirements generally specify a uniform schedule of QC sample analyses. They must be implemented in all laboratories engaged in analysis of compliance samples (e.g., one duplicate sample and one matrix spike sample for every 20 environmental samples). The problem with this approach is that the schedule may provide too few QC samples in laboratories with poor control over imprecision and bias and too many in laboratories with good control over analytical error [Blacker].

This same conclusion was illustrated clearly by Keith et al., using a Radian Corp. expert system named QSA (Quality Sampling and Analysis). QSA calculates the number of QC samples necessary to achieve a given probability of detecting a specified level of bias and precision. Alternatively, it will calculate the probability of detecting bias or precision that exceeds a specified level given the number of QC samples. QSA quickly shows that QA plans that simply rely on a set percentage of QC samples (as almost every one does) provide bias or precision statistics that result in unacceptably low confidence levels [Keith, 1988]. The problem with this standard approach is that laboratories and clients alike are deluded into believing that they are meeting predetermined levels of bias or precision. In reality, they may or may not be achieving the presumed statistical control for precision, because they have not calculated the probabilities of attaining their precision goals; they are only following a traditionally prescribed formula (such as 5% or 10% QC samples).

A novel "intralaboratory" approach has been designed which accounts for the precision and bias of individual laboratories. Through a program of analytical replication and bias adjustment, it permits laboratories with different method performance characteristics to attain user-defined data quality requirements (or acceptable laboratory error rates) [Haeberer, 1990]. Differences due to calibration data bias are accommodated by adjusting laboratory analytical results for matrix spike recovery of the analytes of interest. Differences in imprecision are accommodated by directing laboratories to use laboratory specific levels of analytical replication. Results are then calculated as the average of the replicate analyses. The goal is to control each laboratory contribution to false positive and false negative errors at pre-specified target levels [Blacker]. For pollutant discharge compliance monitoring the acceptable rate of false negative errors would be specified by the EPA. For other types of applications the laboratory client or data user should specify the acceptable rate of both false positives and false negatives.

The intralaboratory procedure uses a laboratory's historical method performance data or a laboratory or carries out a special study to estimate laboratory specific imprecision and recovery of target analytes in the matrix of interest. Using a set of power curves called "replication plan selection curves," the laboratory determines the number of replicate analyses of compliance (environmental) samples and matrix spike analyses needed for its level of imprecision. That way averaged bias-adjusted results do not exceed the target error levels. The laboratory monitors the QC data; if conditions at the laboratory change, they can determine the new numbers of QC samples needed to meet the target levels. The EPA and other clients potentially benefit from the controlled false negatives error rate in the data set, predetermined by the data user. The client determines the false positives error rate in the data set and the expenses associated with controlling it. One of the results of this approach will be that laboratories with good control over bias and imprecision will have to analyze fewer QC samples, and they will be able to offer their services at lower prices than laboratories with poorer imprecision.

QUALITY ASSESSMENT

Quality assessment describes those techniques used to assess the quality of the measurement process and the results. As mentioned above, establishing a system of control charts is a basic principle. After a system has demonstrated statistical control, the charts are used to determine when that system is out of control. Control charts also may be used to visualize or monitor the relative variability of repetitive data, for example, when blanks are used with duplicate or replicate samples, reference materials, and spiked (fortified) samples. Spiked samples, using a similar analyte (surrogate) or the same analyte (standard addition) are used to estimate interference bias when expected confidence limits of percent recovery are exceeded.

A typical control chart is shown in Figure 5. It contains a plot of the data points vs time and a central line that defines the best estimate of whatever response variable is being plotted. Upper (UCL) and lower (LCL) control limits (action limits) define the acceptable bounds within which the plotted values must lie. Usually these are $\pm 3\sigma$ control limits within which 99.7% of the data should lie. Often additional upper (UWL) and lower (LWL) warning limits are also marked on control charts. These are usually ± 2 sigma control limits within which 95% of the data should lie.

Audits should be a feature of all quality assurance programs. A *systems audit* is qualitative and should be made at appropriate intervals to assure that all aspects of the QA program are operative. *Performance audits*, in which a laboratory is evaluated based on the analyses of performance evaluation (PE) samples, are quantitative and also provide valuable quality assessment information. *Data audits* randomly select a few samples from a study. They check all documentation, data entry, calculations, instrument calibrations, data transcription, and report formats for accuracy and conformance to the QA project plan from the time of receipt through the final report [Dupuy, 1990]. Participation in interlaboratory and col-

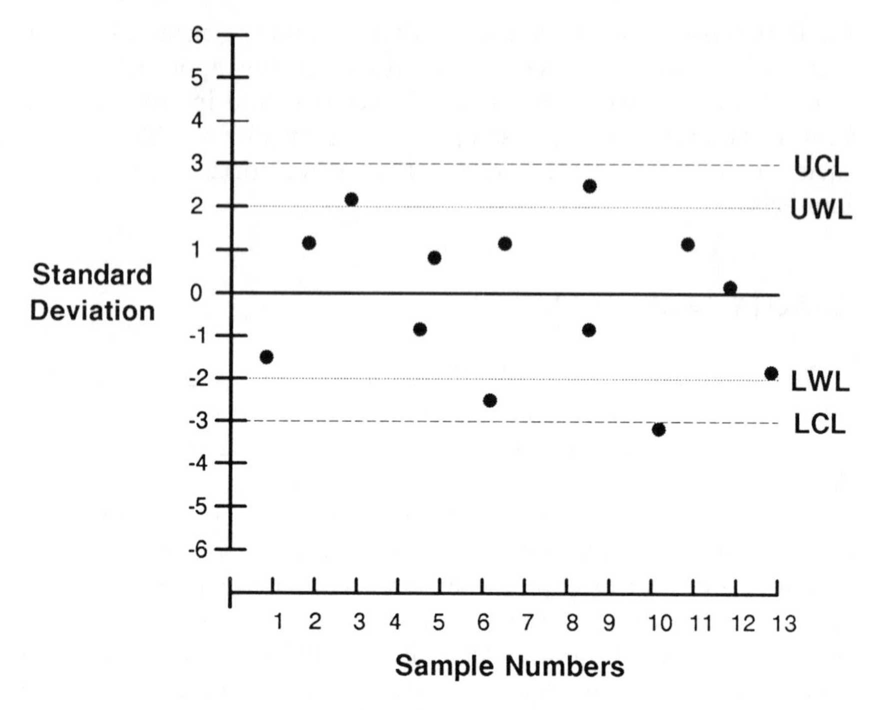

Figure 5. A typical control chart.

laborative test programs is another way for a laboratory to assess the quality of its data output.

MEASUREMENT VARIABILITY

The terms repeatability, reproducibility, intralaboratory variability, and interlaboratory variability are sometimes used to describe various aspects of the measurement process.

- *Repeatability* describes the variation in data generated on a single sample by a single analyst and/or instrument over a short period of time.
- *Reproducibility* refers to variation of data generated over an extended period of time and/or by various analysts or laboratories.
- *Intralaboratory variability* refers to the difference in results when a single laboratory measures portions of a common sample repeatedly.
- *Interlaboratory variability* refers to the difference of results obtained by different laboratories when measuring portions of a common sample.

Because of the nonspecific nature of this terminology, the experimental conditions must be specified whenever such terms are used.

Variabilities due to different operators, equipment, and conditions (but independent of sample variabilities) influence reproducibility and are indeed components of intralaboratory variability. Laboratory biases, when coupled with differences in the capabilities and experience of the analysts at various laboratories, result in an interlaboratory variability that is often very much larger than intralaboratory variability. Well designed and well executed QA programs can help to reduce both intralaboratory and interlaboratory variability [Keith, et al., 1983].

VERIFICATION AND VALIDATION

Verification is the general process used to decide whether a method is capable of producing accurate and reliable data. *Validation* is an experimental process involving external corroboration by other laboratories (internal or external), methods, or reference materials to evaluate the suitability of methodology. Neither principle addresses the relevance, applicability, usefulness, or legality of an environmental measurement. *Confirmation*, a type of verification, is a process used to assure that the analyte in question has been detected and measured acceptably and reliably [Keith, et al., 1983].

The reliability and acceptability of environmental analytical measurements depend upon rigorous completion of all the requirements stipulated in a well defined protocol. A protocol should describe all the necessary requirements of the study including sampling procedures, sample preservation, storage and preparation, analytical measurements, methods, calculations, method verification, validation and/or demonstration of analytical capability, and QA program requirements. The latter may consist of blanks, calibration standards, reference standards, replicate sample analysis, reagent quality control samples, and/or matrix-spiked samples. All analytical results should be critically reviewed. If questions arise during the review, additional confirmatory analyses may be needed, using other methods. In situations where large numbers of samples are analyzed with widely accepted and well documented analytical systems, unusually high results or unexpected low ones, on critical samples, should be checked by a repeat analysis of a duplicate subsample using the same method; if possible, a third subsample should be analyzed by a different analytical method. Agreement of the results indicates the analyte has been measured correctly. Disagreement requires careful study including analysis of additional samples. This approach requires prior planning to collect sufficient amounts of sample at the beginning of the program or to resample if necessary.

Analytical systems that are part of major decision-making processes should use collaboratively studied methods. Confidence in the measurement process is strengthened considerably by participation in the check sample programs conducted by external organizations. Such programs are some of the most effective elements of a QA program.

Qualitative identifications usually should be confirmed. In some cases (for example, monitoring programs) analyte identities are not in question and only require periodic confirmation. Confirmations should be based on a measurement principle or on analytical conditions distinctly different from those used in the initial method. The procedure should be selective and should refer to a different and unambiguous property characteristic of the analytes. Furthermore, the confirmation data must reasonably agree quantitatively as well as qualitatively. Frequently confirmation methods use a second chromatographic column, and identification relies on an analyte signal (peak) within narrow windows of elution time on both columns. If the analyte concentrations calculated from the signal responses on both columns do not agree with one another then the analytes are not the same (i.e., no confirmation—an incorrect identification) or an interfering (false positive) compound is coeluting with the peak exhibiting the larger concentration. In either case the data is invalid and further analysis is required to provide confirmatory evidence. Sometimes this additional analysis is neglected, but logic dictates that it is necessary because the quantitative data, and quite possibly the qualitative data (analyte identification), is in error.

PRECISION AND ACCURACY

Measurements of all types, including quantitative chemical analyses, are characterized by error in the measurement process. This error produces uncertainty in the measurement results. Statistical theory says that this measurement error can be broadly divided into two types: *random error* and *systematic error*.

When we talk about the first of these two errors, random error, we refer to the *precision* of the measurements. Precision describes the degree to which data generated from replicate or repetitive measurements differ from one another. Statistically this concept is referred to as dispersion and is measured using the standard deviation (SD or s or σ) or relative percent difference (RPD) of replicate analyses.

The second component of error in all measurements is systematic error. This error is typically referred to as *bias*. Important sources of systematic error [Hunt and Wilson] include:

- Unrepresentative sampling
- Instability between sampling and analysis
- Measurement inability for all forms of the analyte
- Interference
- Biased calibration
- Biased blank correction

Systematic error is the difference between the limiting mean of a series of measurements and the true value of the property measured [Kirchmer, 1990]. If we were to repeatedly analyze a standard of known concentration, we could estimate the bias in the measurements by comparing the limiting mean (average) analytical result with the true value. This is true because the random errors in the successive measurements would tend to cancel one another out with repetition, and would, by definition, sum to zero over infinite repetitions. The uncertainty in our estimate of bias would depend upon the number of repetitions, since this affects the extent to which the random error is averaged out. The total error in the *average* of repeated measurements depends less and less upon the magnitude of the random error component (imprecision) as the number of measurements increases [D.L. Lewis, 1990].

The meaning of the term *accuracy* is one of the most common points of confusion and ambiguity in discussions of measurement error. There is general agreement that accuracy has to do with "nearness to a true value". Confusion and ambiguity come into the picture when we try to determine "nearness of what?". If we refer to a single measurement result, the nearness of the result to the *true value* (i.e., the accuracy of the result) depends on the magnitude of both random and systematic errors. In this context, accuracy has components of both bias and imprecision. On the other hand, if we refer to the nearness of an *average* value to the true value, we are talking primarily about the bias error component; by taking the average, we have "averaged out" the random errors. Sometimes, accuracy is used in this context, taking on a meaning equivalent to that of the term *bias* [D.L. Lewis, 1990]. This concept is illustrated in Figure 6.

Unless the true value is known, or can be assumed, bias cannot be evaluated. Bias can be estimated only from the averaged measure-

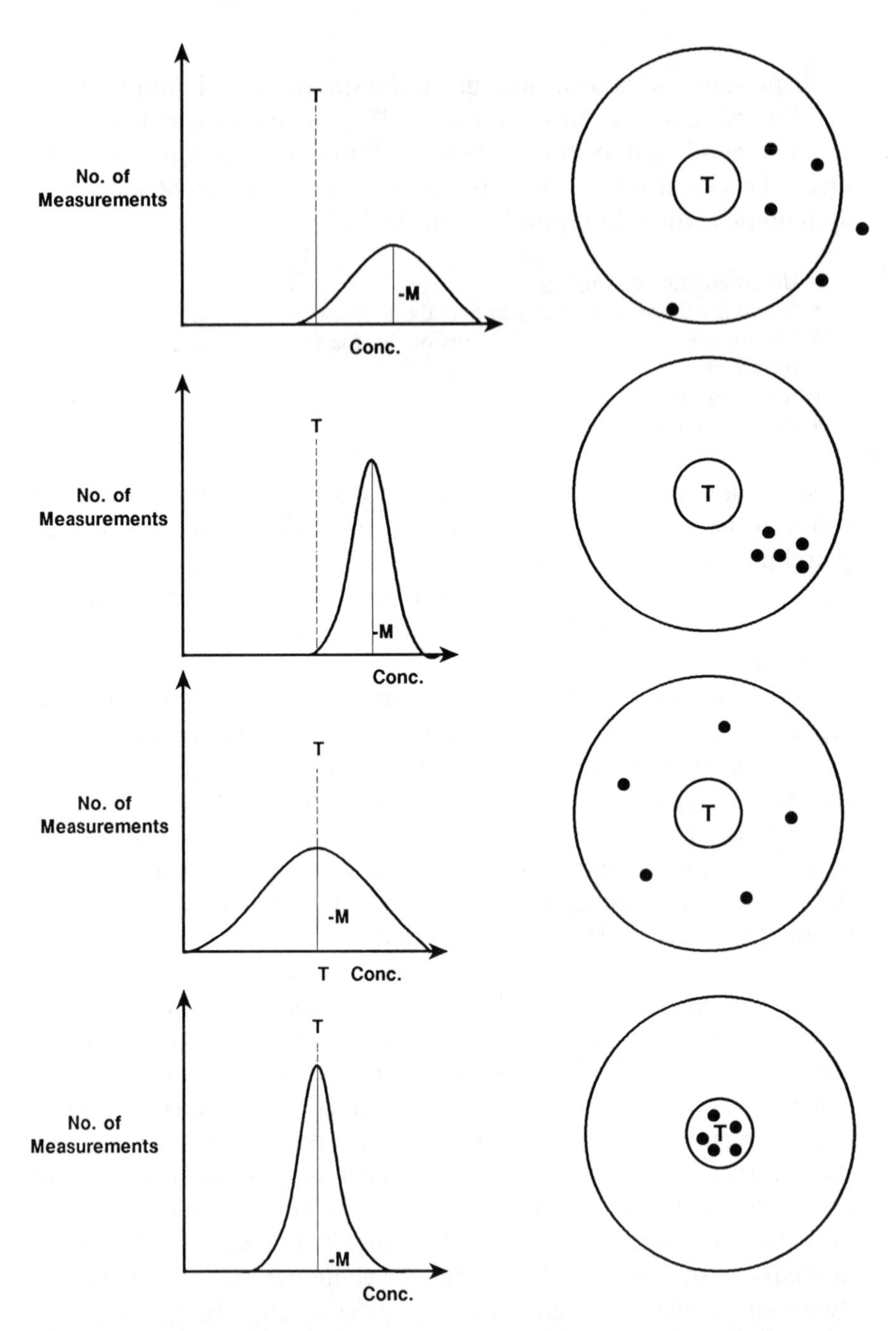

Figure 6. Examples from top to bottom of (1) low precision (large random errors) and low accurancy (large systematic errors), (2) high precision (small random errors) and low accuracy (large systematic errors), (3) low precision (large random errors) and high accuracy (small systematic errors), (4) high precision (small random errors) and high accuracy (small systematic errors), using the traditional bullseye target and Gaussian curve for normal distribution of values. T = true value, M = mean.

ments of samples with known composition. Standard reference materials, when available, are ideal for such an evaluation. Often bias is estimated from the recovery of laboratory-spiked samples. Such spiked (fortified) samples may not fully simulate natural samples so the recovery information should be interpreted with this in mind. Often analyte recovery in spiked samples with complex matrices is biased higher than the recovery of these analytes in the test sample. This bias affects both accuracy and precision calculations.

Despite validated methods, good laboratory practices, and systematic QA procedures, outlier data points can frequently appear in sets of analytical measurements. Outliers are analytical results that differ so much from the average as to be highly improbable. Statistical techniques may be used for their identification [Barnett and Lewis]. Zero or negative results are often considered to be outliers, but when working near the limit of detection, a certain number of analyses are expected to be zero or negative by chance alone [Rogers]. When any outliers are discarded from a data set, it is important that they be identified and that the statistical or operational reasons for their deletion are given.

CHAPTER 9

Analytical Measurements

Measurement errors are of three types. Systematic errors are always the same sign and magnitude, and they produce biases. They are constant no matter how many measurements are made. Random errors vary in sign and magnitude; they are unpredictable and average out and approach zero if enough measurements are made. *Blunders* (a type of systematic error) are simply mistakes that occur on occasion and produce erroneous results. Measuring the wrong sample, transcribing or transposing measured values, misreading a scale, and losing due to mechanical error, are examples of blunders. They produce outlying results that may be recognized as such by statistical procedures, but they cannot be treated by statistics. Appropriate quality control procedures can minimize some kinds of blunders, but they may not eliminate carelessness, which is the principal cause. In addition, there can be random components of systematic errors. For example, an instrument zero reading is adjusted to eliminate a systematic error, but this adjustment has a certain amount of imprecision. Any single adjustment of a zero will produce a bias; however, over many adjustments the imprecision should cancel out. The uncertainty of this adjustment can be the imprecision of the setting process [Taylor, 1987].

Analytical measurements should always be made with properly tested and documented procedures. Furthermore, every laboratory and analyst must conduct sufficient preliminary tests, with the methodology and typical samples, to demonstrate competence in the use of the measurement procedure [Taylor, 1983]. The analytical procedures should include quality control samples and calibration curves so that random and systematic errors will be minimized. The analytical procedures should provide the required precision, minimum artifact contamination, and best recovery possible. Contamination can be introduced from sampling containers, equipment, reagents, solvents, glassware, the atmosphere and added surrogates, or internal standards. Keeping the number and complexity of operations to a minimum will lessen contamination possibilities from these sources.

To establish confidence in the analytical measurements, sensitivity, accuracy, precision, and recovery data should be comparable, when possible, to the values of similar samples in literature. State of the art analytical techniques are not always necessary, but the results and methods used should be able to stand the test of peer review.

DEFINITION OF THE DATA SET

Quality control data should be obtained from several types of QC samples, in addition to the environmental samples. Common QC samples include replicate environmental samples, trip blanks, reagent blanks, background (control) samples, calibration standards and calibration check standards, spiked field or laboratory blanks, spiked environmental samples, and reference (quality control check) standards.

- *Environmental sample data* are the objectives of an investigation.
- *Replicate environmental samples* measure the overall precision of the sampling or analytical methods.
- *Replicate analyses* are identical analyses carried out on the same sample multiple times. They measure laboratory analytical precision only.
- *Trip blanks* determine if interference or contaminants are introduced during the entire process of collecting, shipping, and storing samples. These were also discussed in Section I.
- *Matrix spiked laboratory blanks* measure the normal level bias due to matrix effects and analytical method errors, including laboratory contamination, calibration errors, etc. They consist of solvent or reagent blanks spiked with the analytes of interest.
- *Spiked (fortified) field blanks* and spiked environmental samples measure the effects the sample matrix may have on the analytical methods (usually analyte recovery).
- *Laboratory blanks* (also called background matrix samples) are matrices, without the analytes of interest, that are carried through all steps of the analytical procedure. All reagents, glassware, preparations, and instrumental analyses are included [Kulkarni]. Solvent blanks, reagent blanks, and instrument blanks are types of laboratory blanks. They are used to measure

contamination when stirring, blending, digesting, or subsampling and to prepare samples prior to analysis. Usually at least one laboratory blank is prepared for every 20 environmental samples processed [Black]. This number seems to be an arbitrary convention and its deficiencies were pointed out in Chapter 8.

- *Solvent blanks* consist only of the solvent used to dilute or extract the sample. They identify and/or correct for signals produced by the solvent or impurities in the solvent [D.L. Lewis, 1988].
- *Reagant blanks* consist only of the reagents used to prepare the samples, and they are analyzed the same way as the test samples. They identify and/or correct the signals produced by the reagents or imurities in the reagents.
- *Instrument blanks* (also called system blanks) are solvent or reagent blanks that measure interference or contamination from an analytical instrument. They cycle matrices, containing materials that are normal to the analysis but minus the analytes of interest, through the instrument [Kulkarni].

Two types of standards are generally used to determine whether an analytical procedure is in control: calibration check standards (also called calibration control standards) and laboratory control standards. Calibration check standards are solutions containing the analyte(s) of interest at comparable (often low) but known and measurable concentrations [Black]. Usually these standards consist of the analytes of interest at 3 to 5 different concentration levels that bracket the lower and upper expected concentration levels. Calibration check samples are used as the basis for quantitating the analytes in test samples. They are also used to assure that calibration stays within control and that the detection limit is acceptable [Meier]. Laboratory control standards (QC check standards) require certified standards, generally supplied by an outside source [Black]. They are used to measure normal level bias originating from procedural or operator errors or contamination from laboratory sources.

Blanks are important to all environmental analyses and become critical as the detection levels of the analytes of interest are approached. Detection levels should be determined with the variability of a blank. When an analysis involves a peak rising above a background and no true blank response is obtained, an alternative procedure must be used. This may be done by determining the method detection level (MDL) from the standard deviation of low level standard responses, or by establishing some concentration that must be exceeded based on standard responses [Kirchmer, 1988].

Generally, EPA protocols do not permit you to subtract blank values from sample values in their final reported form. However, subtracting blank values from data used to determine limits of detection and quantitation is a different philosophy, and it should always be performed in order to establish these limits accurately. Levels of detection and quantitation can contain significant errors if blank values of zero are used when they are not zero [Kirchmer, 1988].

The frequency and order for measuring a sequence of these various QC samples and blanks should be defined in the protocols developed in the program planning stage. Many laboratories have adopted the practical approach of analyzing groups of samples, using the same analytical method as sets. A set consists of a finite number of environmental samples plus a predetermined number of various quality control samples (blanks, QC check samples, replicates, etc.). All the test samples and QC samples in the set are prepared and analyzed together under the same conditions (preferably including the same personnel). The QC samples assess all the test samples within that set. Thus, the set, which may include environmental samples from more than one client or project, stands on its own merits. If the QC samples in the set are unacceptable for any reason, only the environmental samples in that set must be reanalyzed [Dupuy, 1990].

CALIBRATION AND STANDARDIZATION

Calibration determines the correctness of the assigned values of the physical standards or the scales of the measuring instruments. The whole purpose of calibration is to minimize bias in the measurement process. Typical calibrations include standards for mass, volume, and length; instruments that measure temperature, pH, and chemical composition also can be calibrated. The term *standardization* is used frequently to describe the determination of analytical instruments response function, but it should refer to the entire method [Kirchmer, 1990].

Calibration accuracy critically depends on the reliability of the required intercomparison standards. Likewise, chemical calibration or standardization critically depends upon the quality of the chemicals in the necessary standard solutions and the care exercised in their preparation. Another important factor is the stability of these standards once they are prepared. Calibration check standards are prepared fresh periodically usually from neat chemicals, and used to verify the validity of the calibration curve. Of course, the frequency with which these standards are prepared depends on their stability. Calibration check standards should be freshly prepared every three months for stable analytes, and more frequently (weekly or daily) for less stable analytes. Generally, calibration check standards should be analyzed at the beginning of each day analyses are performed, if the analytical system is not newly calibrated. These standards may be internal or external; however, if they are external standards a second analysis should be performed at the end of the day. Every standard should have both the date of preparation and an assigned expiration date printed clearly on its label. It is also a good idea to put the name or initials of the preparer and the notebook page number on the label.

Calibration uncertainty may be characterized according to the confidence in the standards and the uncertainties of their use in the measurement process. It is important to be able to trace calibration standards back to an acceptable reference standard (for example, EPA or NIST standards and vendor-certified standards). The uncertainty of the chemical standard composition will depend on the degree of experimental realization of the calculated composition. This is based on the knowledge of the constituent purity, the accuracy of the preparative process, and considerations of stability. The reliability of transferring the standard to the calibrated system is a further consideration. Both systematic and random sources of error are involved in all of the above and will need to be minimized to meet the accuracy requirements of the data. Repetitive calibrations will decrease the random component of uncertainty but not any biases. As calibration uncertainty and measurement uncertainty approach each other, calibration can become a major activity, even in routine measurements [Taylor, 1987].

A calibration line may show the statistical uncertainty of fit (Figure 7) as a band surrounding the fitted line. In the case of equispaced data, the band is narrowest at the center of the plot and broadens at the extremities. The center has this increased confidence because it has the highest number of degrees of freedom; the number decreases in each direction from the center. The lower confidence when using such a line in an extrapolation mode is a consequence of this as well as the concern for linear calibration data over an extended range. The figure shows qualitatively how the uncertainty of the calibration line adds to measurement uncertainty each time it is used [Taylor, 1987]. However, the imprecision of the calibration curve has little effect on the confidence limits of an individual analytical result [Cheeseman and Wilson]. The practical result is that confidence limits of most analytical results are primarily influenced by random error in the signal responses of the measurement process.

Usually at least three different concentrations of calibration stan-

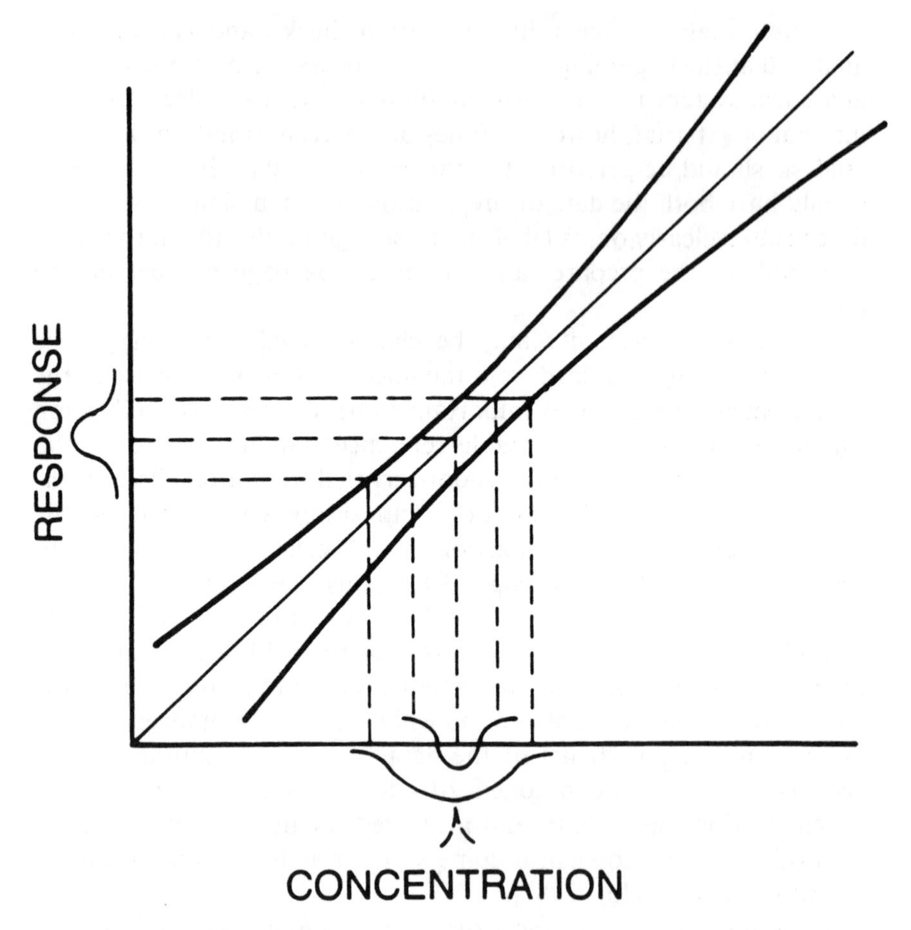

Figure 7. Propagation of calibration uncertainty. The uncertainty of the calibration line is shown as a band of varying width. This uncertainty adds (in quadrature) to measurement uncertainty when considering the total analytical uncertainty. Reprinted from *Quality Assurance of Chemical Measurement* by J. K. Taylor, 1987, p. 104, with permission from Lewis Publishers.

dards should be measured (EPA sometimes specifies five). The concentrations of the calibration standards must bracket the expected analyte concentration in the samples. The concentration data, when plotted graphically, are referred to as a calibration curve. Analytes with concentration levels outside calibration curve limits can only have estimated concentrations, since calibration curves may not be linear above or below the concentration levels used. An important consideration in preparing calibration curves is that the calibration must be done under the same instrumental and sample conditions as those that will exist during the measurement process. Calibration frequency depends on the DQOs of the investigation and the stability of the measurement system.

Internal standardization involves the addition of a reference material to the sample. *External standardization* involves separate use of a reference material. The internal standard material is chosen to simulate the analyte of interest; the external standard material is usually the analyte being measured. The ratio of the internal standard response to the analyte response is called the relative response factor, and it calculates analyte concentration.

Another technique commonly used is *standard addition* (spiking), where successive, increasing known amounts of analytes are added to the sample or aliquots of it. Standard additions correct for interferences which are proportional to environmental sample concentration. This situation occurs when the slope of the calibration curve is different for the analytes in the standards and the environmental samples [Kirchmer, 1990].

Surrogates are check standards added to every sample in known amounts at the beginning of an analysis; they are not one of the target analytes and not expected to be found in the environmental samples. In addition, surrogates must be chemically close to the analytes of concern [Kirchmer, 1990]. Surrogates are sometimes incorrectly termed internal standards; however, surrogates may be used as spikes in a technique known as "surrogate spiking". The difference is that an internal standard is the reference material against which the analyte signal is directly compared, whereas the signal from a surrogate is not used directly for quantitation. Both types of reference materials are chosen to simulate the analyte of interest. Surrogates may be used indirectly for quantitation as, for example, in determining the recovery efficiency during sample pretreatment. In each of these techniques, the spiked sample is analyzed under the exact same conditions as the actual sample, including all phases of pretreatment. Recoveries of the analytes of interest are inferred from those found when using spiked materials.

Isotope dilution analysis is a well known technique used with elemental analyses that has recently been perfected with organic compounds by mass spectrometry in complex environmental matrices. The technique involves spiking stable-labeled isotopes (for example, deuterium labeled) of each analyte of interest into the sample prior to analysis (i.e., before extraction or any other pretreatment). The spiked matrices are quantitated with stable-labeled isotopes as internal standards. By assuming that recoveries of the environmental sample analytes will be the same as their spiked isotope analogs, and that the response of recovered spikes are equal to the amounts added, the recovery of the stable-labeled isotopes can be determined by external standard calibration.

It is essential with any of these techniques to demonstrate that the spiking procedure is valid. It must be shown either that the spiked chemicals equilibrate with the corresponding endogenous ones or that the recovery of the spiked chemicals is the same as the recovery of the endogenous chemicals. It must be within experimental error and over the full range of concentration levels to be analyzed [McKinney].

PREPARATION OF SAMPLES

After a sample has been obtained, the analytical protocol often requires one or more treatments prior to actual analyte measurement. Sample preparation may involve physical operations such as sieving, blending, crushing, and drying and/or chemical operations such as dissolution, extraction, digestion, fractionation, derivatization, pH adjustment, and the addition of preservatives, standards, or other materials. These physical and chemical treatments not only add complexity to the analytical process, but they are potential sources of bias, variance, contamination, and mechanical loss. Therefore, sample preparation should be planned carefully and documented in sufficient detail to provide a complete record of the sample history. Further, samples taken specifically to test the quality control system (i.e., quality control samples) should be subjected to these same preparation steps [Keith, et al., 1983].

The analyst must recognize and be aware of the risks that are associated with each form of pretreatment and should take appropriate preventive action for each. This may include reporting, correcting for, or possibly removing interferences from the analytes of interest by modifying the analytical protocol. All changes in an analytical protocol must be documented, validated, and tested at the same level as the original method. Sometimes the only way of finding an unexpected analytical result lies in the sampling documentation or analytical protocol changes, which were necessary at the time for some logical and specific reason.

RECOVERY

Analyte recovery is influenced by such factors as analyte concentration, sample matrix, and storage time. Because recovery often varies with concentration, the spike and the analyte concentrations should be as close as practical. This is not quite as important when isotopically labeled standards are spiked for isotope dilution analysis, because the methodology automatically corrects for sample matrix recovery effects. It is not unusual to find significant recovery differences for organic standards spiked into industrial wastewater samples taken only days (or even hours) apart; the composition of wastewater samples often changes significantly with time. This variability is one reason why isotope dilution analyses are becoming more important with complex wastewater analyses.

Another consideration in evaluating recovery efficiencies is the amount of time a sample has been stored before pretreatment and analysis. Analyzing a sample that has been stored for a long period of time sometimes will give different values for the analytes than when the sample is analyzed fresh. This difference may be caused by changes in the matrix and/or the analytes. Therefore, if it is anticipated that a sample is going to be stored for an appreciable period of time before analysis, it should be demonstrated that significant changes have not occurred. What an "appreciable period of time" is depends on the analytes, sample matrix, and other chemical and physical factors determined in the planning phase. When EPA or other validated methods are used, the *maximum holding time* for a sample prior to its analysis (or pretreatment) is usually a clearly defined limit in the method. An appropriately designed quality assurance program can help determine the potential effects of storage time on analyte loss, and it will specify maximum holding times as necessary if they are not specified by the analysis method.

As mentioned above, variable recovery is sometimes resolved by using the isotope dilution technique in which isotopes of the analytes being measured are spiked directly into the sample. Both radioactively and nonradioactively labeled compounds or elements can be used. This principle is based on the assumption that the labeled compound or element will behave in an essentially identical manner to the unlabeled analyte of interest during sample preparation.

INTERFERENCES

Because of the complexity of environmental samples and the limited selectivity of most methodologies, interference is common. Appropriate quality control samples and experiments must be included to verify that interferences are not present with the analytes of interest or, if they are, that they be removed or accommodated. Interferences that are not accommodated cause incorrect analytical results, including false positives and false negatives.

Interferences arise from two sources that may occur simultaneously: (1) constituents that are inherent in the sample and (2) contaminants or artifacts that have been introduced during the sample collection and/or analytical process. A good sampling and analysis plan and quality control program, incorporating an appropriate experimental design with matrix and method blanks, is critical to identifying the sources of interferences in test samples and differentiating between interferences and artifacts. Large interferences may sometimes be avoided by changing to different methods. Contaminants (artifacts) are avoided or identified by an appropriate, comprehensive quality control program.

CHAPTER 10

Documentation and Reporting

Documentation and reporting is one of the most controversial subjects today in environmental analytical chemistry, because it affects how data are received and perceived by the user, and often the public at large. The data included in reports, as well as the data not included, is a sensitive issue.

There are several reasons for this controversy. Some are technical and others fall more in the sociopolitical-economic realm. It is this latter interest, of course, that ultimately provides the driving force behind the concern.

Analytical chemists must always emphasize to the public, that the single most important characteristic of any result obtained from one or more analytical measurement is an adequate statement of its uncertainty interval [Rogers]. Lawyers usually attempt to dispense with uncertainty and try to obtain unequivocal statements; therefore, an uncertainty interval must be clearly defined in cases involving litigation and/or enforcement proceedings. Otherwise, a value of 1.001 without a specified uncertainty may be viewed as legally exceeding a permissible level of 1 [Keith, et al., 1983]. While that is ridiculous, it may nevertheless carry legal consequences.

DOCUMENTATION

Documentation of analytical measurements should provide information sufficient to support all claims made for the results. Documentation requires all the information necessary to (1) trace a test sample from the field to the final results, (2) describe the sampling and analytical methodology, (3) describe all confirmatory evidence, (4) support statements about levels of detection and quantitation, (5) describe the QA program and demonstrate adherence to it, and (6) support confidence statements for the data [Keith, et al., 1983].

Data, including all instrumental output, must be recorded in laboratory notebooks or other suitable media (for example computerized magnetic media). It should include complete sample documentation, transfers and movement (including when, where, and by whom the sample was received, prepared, and analyzed), sample number, initial sample weight, extraction volume, final weight and volume analyzed, instrument response, any spike added, sample calculations, and sample concentration as appropriate. The time when spiked samples and blanks were analyzed relative to test sample measurement is of the utmost importance to data evaluation.

A good practice is to archive all records within a completed project in a room or container that will protect them from fire or water damage. Logbooks, chromatograms, control charts, laboratory notebooks, final laboratory reports, and pertinent correspondence are examples of the data that should be archived. Laboratory records should be kept in a permanent file for a length of time set by the government, other legal requirements, or the employing institution, whichever is longer.

Another good practice is to make frequent temporary copies of digital data from analytical instruments such as mass spectrometers, gas chromatographs, etc. Temporary copies on magnetic tape or diskettes, stored in a location separate from the laboratory, provide cheap insurance against the loss of current data from fire or any other accidents.

Bound notebooks are preferred to looseleaf-type notebooks [Keith, et al., 1983]. Whenever legal or regulatory objectives are involved, the notebook data should be entered in ink, each page should be signed and witnessed, and all errors or changes should be struck through one time (so they are still readable) and initialed. In fact, other than a witness signature, the above should be a standard operating procedure for entering all data all the time.

DATA GENERATION VS DATA USE

Laboratories generate and perform QC checks on individual measurements; they are reported as individual analytical results and associated QC data. However, users usually compile these individual measurements into *sets of data*, and reports and conclusions are generally made from these sets of data. Therefore, the laboratory is responsible for producing individual test measurements with analytical systems that are in statistical control and reporting that data with a statement of its uncertainty interval. This means providing appropriate rounded or truncated data that have a specified uncertainty interval (+ or - some percentage). Uncertainty intervals may be quoted for an individual analyte, or more often, for a specific method as required in many of the EPA methods currently promulgated. Laboratories have the responsibility to provide this information with every analytical report.

The data user should request these uncertainty intervals from the laboratory if they are not provided, because *the responsibility for using and presenting final data belongs with the user and not the laboratory.* The user should seek help from the laboratory or another source to determine what data to present in a report, but the laboratory is not responsible for deciding whether or not go give the user censored reports; the user should request censored reports if these are desired.

REPORTING BY LABORATORIES

There are many terms that are commonly used when describing low levels of analyses. The three that have been selected for this discussion are: limit of detection, method detection limit, and instrument detection limit. Each one is quite different from the other.

- *Limit of Detection* (LOD)—the lowest concentration level that can be determined statistically different from a blank at a specified level of confidence. This corresponds to the "critical level" [Currie, 1988].
- *Method Detection Limit* (MDL)—the minimum concentration of a substance that can be measured and reported with 99% confidence that the analyte concentration is greater than zero. It is determined from analysis of a sample in a given matrix containing the analyte [U.S. EPA, 1984]. The MDL considers all of the analytical operations on a sample (extractions, concentrations, reagents, etc.). It is usually a preferred term used by the EPA. The MDL corresponds to the "criterion of detection" [ASTM].
- *Instrument Detection Limit* (IDL)—the smallest signal above background noise that an instrument can detect reliably. Because the IDL only involves

one component of the analytical process, it is seldom used with environmental analyses. The IDL must not be confused with the average values from a blank sample, since the blanks may incorporate important bias effects from the environmental sample matrix and/or sample preparation and handling.

Detection of an identified analyte is the most important decision in low-level analysis. One of the first questions that must be answered is whether the analyte of interest is present in the sample. A major problem with specifying the LOD is that it is a calculated concentration level that is indirectly *selected.* The concentration level of the LOD is calculated from the risks of false positives and false negatives. Unfortunately, the criteria used for these risk selections are not always understood. Furthermore, as will be shown later, the selected value that determines if an analyte is reported present in a sample may be different from the selected value that determines if an analyte is not reported as present, and that can be confusing. To paraphrase (badly) Shakespeare, to report or not to report . . . that is the question!

The question of when and how to report the presence of an analyte at or near its detection limit is one that is still the subject of disagreement. There are basically two opposing camps: one advocates reporting all values without regard to concentration and the other opposes the reporting of any values below the LOD or MDL. One solution to this dilemma is to separate analytical low level data reporting rules into separate and distinct protocols: one set of protocols is for laboratories that generate the data and must not bias them, and a second set of protocols is for the data user who must interpret the individual data values as sets of data with limitations and uncertainties.

In the American Chemical Society-sponsored publication entitled *Principles of Environmental Analysis*, [Keith, et al., 1983], the LOD was defined as an analyte signal that is three times the standard deviation (3σ) of its measurements above the corresponding well characterized blank response. Furthermore, it was recommended that analyte values below this *selected LOD* should be reported as "not detected" (ND) and that the limit of detection (LOD) should be given in parentheses: ND (LOD = value).

The reason for the above recommendations was that published or reported analytical data are often used in contexts that extend beyond the originally defined boundaries. Furthermore, published or reported data may be incorporated as portions of other data compilations, computer databases, or summary reports. Process notations, explanations, footnotes and similar qualifications, and limitations associated with the data are often deleted. When this happens the data are likely to be used or interpreted erroneously with assumptions of validity beyond their intended use. Although there is a greater awareness of the specific limitations under which most environmental analytical data sets are generated today, the propensity for misuse of environmental data still exists.

Even though most environmental analytical laboratories and their clients have embraced the above ACS recommendation for reporting analytical data, it has not been universally accepted by all chemists and statisticians. The practice of not reporting data below a 3σ LOD has sometimes been described as "censoring of results to prevent a possible faulty inference being drawn . . . " [ASTM], data "censored on the left" [Gilbert], and "left-censoring" [Newman, et al.]. This terminology refers to the discarding of all values less than the MDL, LOD, or other designated reporting limit when ND is used for values below the LOD under the usual practices. Important information about the data variability at these low levels is lost with this convention. The key issue with respect to recommending that results below the LOD be reported is one of obtaining accurate means or averages of results that do not depend on any assumptions [Kirchmer, 1990].

When is an Analyte Detected?

To understand the above criticism concerning censored data, and thus to intelligently address the selection and use of LOD, we must recognize that there are really two fundamental issues that define whether an analyte is or is not detected. Furthermore, we must understand that the usual practice of chemists is to merge these two issues into one, leading to criticisms based on theoretically and statistically correct concepts but often different from actual laboratory practice. A good presentation of this problem is provided in an overview by Currie [Currie, 1988].

The first fundamental detection issue involves a qualitative, binary (yes or no) decision as to whether an identified analyte has been detected. The measurements used in chemical characterization are not direct measurements of concentration, rather they are derivations of a known relationship between the magnitude of an instrument signal and the concentration of the analyte(s) of interest. Some response from instrument "noise", sample preparation contamination, etc., is always included in the measurement system. This noise interferes with that portion of the signal attributable to the analyte and thus with determination of the analyte concentration in the sample. Deciding, based on an observed signal, whether or not the blank corrected signal actually indicates the presence of analyte is the first issue in detection. This involves distinguishing a "real" signal from the blank response signal, and it requires establishing a criterion or decision rule to classify an observed signal as detected or not detected.

The first detection issue represents the binary decision process characteristic of statistical hypothesis testing. As a binary decision, there are only two possible choices, and one must be selected. For each choice, in a given situation, the decision will either be correct or incorrect. The decision process is subject to the two types of measurement errors characteristic of hypothesis testing. One type is the error of concluding that the analyte of interest is present when, in fact, it is not. This is a Type I error and the risk of a Type I error is denoted by the Greek letter alpha (α). Type I errors are also commonly known as *false positives.* This type of error can only result from a decision of "detected", and it is equivalent to incorrectly rejecting the null hypothesis in an hypothesis test. The second type of error is the error of concluding that the analyte of interest is not present in the sample when, in fact, it is. This is a Type II error and the risk of a Type II error is denoted by the Greek letter beta (β). Type II errors are also commonly known as *false negatives.* This type of error can only result from a decision of "not detected", and it is equivalent to incorrectly failing to reject the null hypothesis in a statistical hypothesis test. The decision process and these two types of measurement error are illustrated in the form of a decision table in Figure 8.

The risk or probability of the first type of error, false positive, is determined by the choice of the decision point, or decision criterion, which has been referred to as the LOD. The LOD represents a criterion for detection decision, i.e., deciding whether to classify a result as detected or not detected when the observed signal is close to that obtained for blanks measurements (i.e., similar to background noise). The MDL is also a criterion for making a detection decision, but zero concentration is used instead of the signal response from the blank. When the blank has zero concentration of the analytes of interest, the MDL and the LOD are the same. The LOD or MDL represents the signal (e.g., chromatographic peak area), which yields a positive detection decision. These signals depend on analyte concentration and LOD or MDL signals or greater lead to a detection decision of "detected"; signals below this threshold level result in a decision of "not detected". By establishing this threshold level a safe distance (3σ) above the average signal which is generated in the

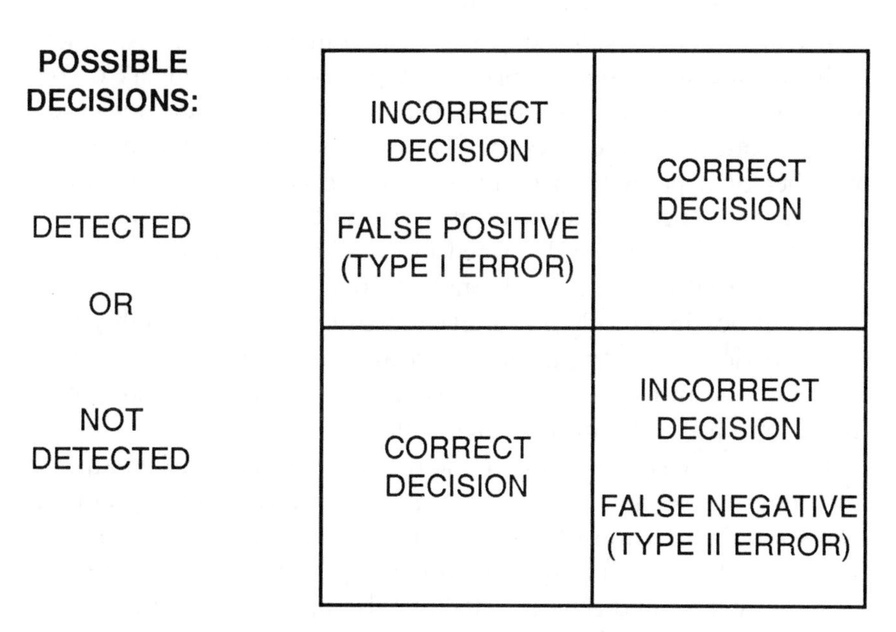

Figure 8. Decision table illustrating detection decision (LOD).

absence of an analyte (MDL) or for a blank (LOD), the chance of making a false positive decision is minimized.

When is an Analyte Reliably Detected?

The first fundamental decision (LOD or MDL) represents the concentration level at which the user can decide if an observed signal represents the presence of an analyte of interest. The second fundamental decision represents the concentration level at which detection is extremely likely. In other words, it represents a level at which there is very little chance of failing to detect an analyte if it is present. At the LOD or MDL there is a 50% chance of detecting an analyte and a 50% chance of missing it (assuming measurement error is symmetrically distributed).

The concentration level at which a detection decision is extremely likely to be made correctly has been called the detection limit (DL)

[Currie, 1968 & 1988]. However, this is so close in terminology to limit of detection that it may be confusing; a more descriptive term, *reliable detection level* (RDL) will be used in this discussion. What led to the criticisms of MDL or LOD mentioned earlier is the common practice of assuming that the RDL and MDL or LOD are equivalent. This assumption is not usually correct, as will be explained in the following discussions.

Because there is some degree of random error and instrument noise in any measurement process, repeated measurements of the same property, under the same or similar conditions, do not necessarily produce identical results. Thus any single result is only one member of a population of many possible results which could be obtained if repeated measurements are made. This distribution of possible results may be described in terms of a *probability density function* (PDF). The PDF defines the relative probabilities of the different possible results for a given measurement. Probability density functions representing low-level analytical measurements help illustrate the false positive and false negative risks in working near the lower end of the measurement range.

Consider the situation where the true concentration of an analyte of interest in a sample is zero. A PDF curve representing normal Gaussian distribution of possible measurement results is shown in Figure 9. There is an equal theoretical probability of positive or negative signals and thus the risk of false positives (α errors to the right of zero) and false negatives (β errors to the left of zero).

Now consider the usual situation where the MDL and RDL are not distinguished from one another and exist at some value above zero. Figure 10 shows a PDF curve for results of possible measurements made on a sample at three standard deviations above zero background. In this figure the PDF results for a sample having zero concentration are the same as those in Figure 9, and the level of detection and the reliable detection level are the same. The risk of false positives at 3σ from zero is less than 1%; this α error is shown as the small black area to the right of the concentration level at 3σ (not drawn to scale). Note that 50% of the results would fall below the MDL (the shaded area) and thus be labeled "not detected"; therefore, an analyte at this concentration has only a 50% chance of being detected. Statistically half of the data will be discarded as "not detected" under these conditions. This is the basis for the criticism that the selection of 3σ as the MDL or LOD produces left censored data. However, when the analyte is present at this concentration it

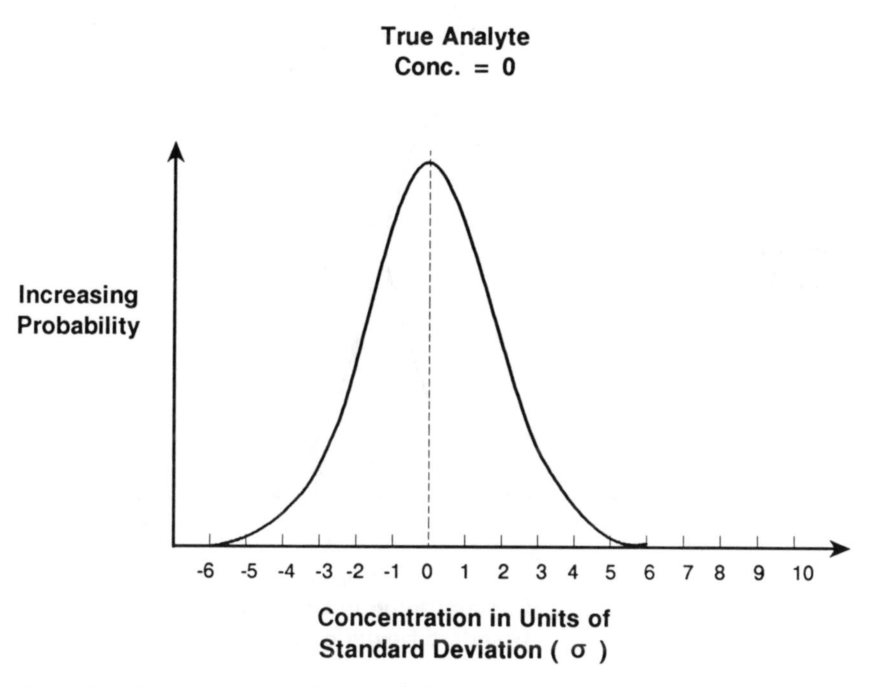

Figure 9. Graphic representation of a PDF curve, where the true value of an analyte is zero.

will be detected only half the time and this 50% risk of a false negative is too high to be considered *reliable detection.*

One solution to this criticism is to establish the reliable detection level (RDL) at a different, higher concentration, where the risk of a false negative (β error) becomes acceptably low. Other authors (Currie, 1988, Cheeseman and Wilson, and Hunt and Wilson) have proposed this same solution but used different terms to describe it. This is depicted in Figure 11 where the false negative risk becomes less than 1% because the RDL is twice the MDL. In this example both false positive risk (solid dark area to the right of MDL) and false negative risk (solid dark area to the left of MDL) are less than 1% (not drawn to scale).

The key to selecting the risks of errors from which the MDL and RDL are calculated lies in having, or obtaining, some information about the random variability, or imprecision, in the measurements of interest. This random variability is typically expressed in terms of the standard deviation. For a very reliable measurement process characterized by a normal (i.e., Gaussian) distribution, approximately 99.7% (99.72%) of the individual measurements should fall within plus or minus 3σ of the mean. The remaining 0.28% fall outside

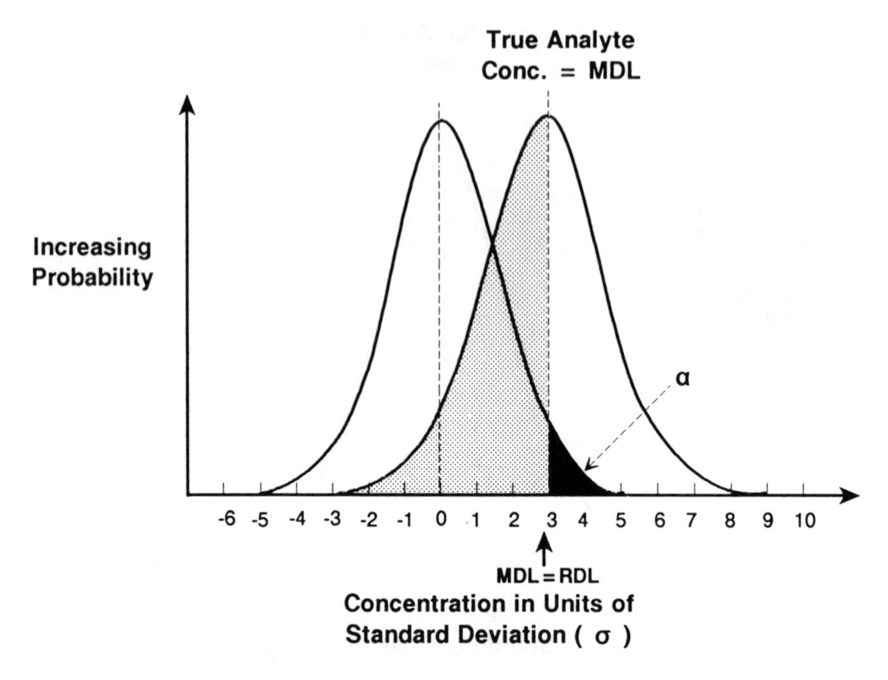

Figure 10. Graphic representation of an unbalanced false positive/false negative risk when the MDL and the RDL are the same.

these limits, with half in each of the two tails of the distribution as shown in Figure 11. For an unbiased analytical method, repeated measurements of a blank sample should give a mean measured concentration of zero; thus, it should be possible to define the MDL of such a measurement process as a concentration 3σ above zero, with the false positive risk set at "less than 1%" (a theoritical risk of 0.14%). For an equivalent false negative risk, the RDL for this hypothetical measurement system would be at a concentration another 3σ higher (twice the MDL or 6σ).

If a higher degree of false positive or negative risk is acceptable, then lower concentration levels may be *selected*. For example, if a 1% risk of false positive and false negative errors is acceptable, then the MDL may be 2.33σ above the blank, and the RDL would be twice that or 4.66σ. Thus, the RDL at 4.663σ would be the lowest analyte concentration that can be reliably detected with a 99% confidence level. Or, the RDL at 6σ would be the lowest analyte concentration that can be reliably detected with a 99.9% confidence level. A range of false positive and false negative risks is summarized in Table 3. Notice that the key word is *selected*; someone must select the risk levels [alpha and beta]. The MDL, LOD, and/or RDL are deter-

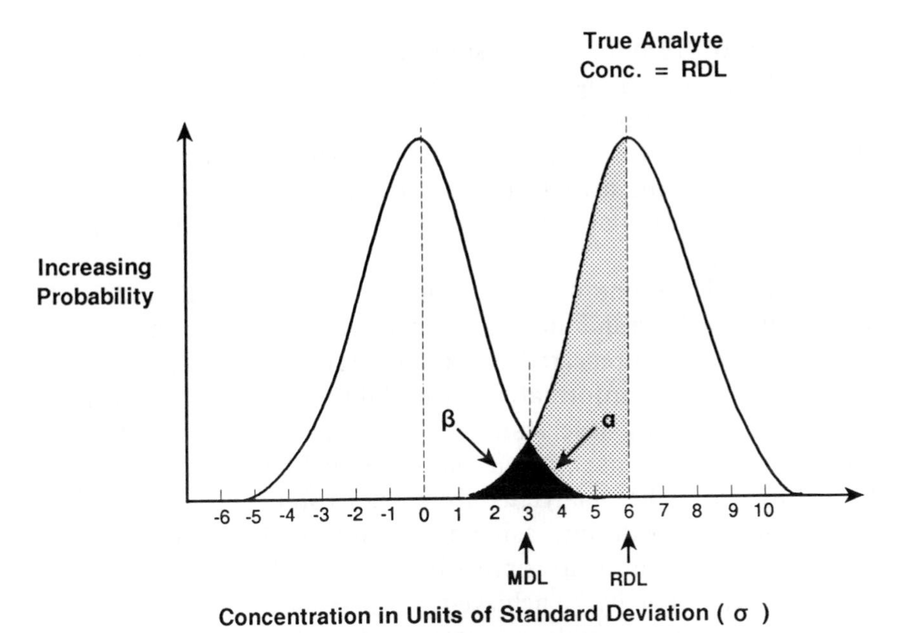

Figure 11. Graphic representation of the MDL and the RDL, where false positive (α) risk and false negative (β) risk are equal, and each is less than 1%.

mined for those levels of risk. Usually, the acceptable risk values are selected by the analytical laboratory or the client.

It should be noted that, in the forgoing discussion and in Table 3, we talk about risks corresponding to different multiples of σ, the population standard deviation. From a strict statistical standpoint, this implies that the true variability of the system of interest is known. In actual practice this is rarely the case. Instead, we usually deal with the sample standard deviation(s), which is estimated from a limited data set. When s is used as an estimate of σ, instead of using the multipliers shown in Table 3 (which represent z values), it would be more correct to multiply the estimates of standard deviation (i.e., s, sample standard deviation) by a Student t value appropriate for

Table 3. Probability of False Positive/False Negative Determinations

MDL or LOD	RDL	False Positives [α]	False Negatives [β]
3σ	3σ	0.1%	50%
3σ	4σ	0.1%	16%
3σ	5σ	0.1%	2.3%
3σ	6σ	0.1%	0.1%
2.33σ	2.33σ	1%	50%
2.33σ	4.66σ	1%	1%
1.64σ	3.28σ	5%	5%

the degrees of freedom of our estimate. This is the basis of the EPA approach to determining MDL [EPA, 1984]. Since this is only one of several potential theoretical limitations to this approach, it may be sufficient, for most purposes, to simply recognize this limitation and understand that the actual probabilistic risks of false positive and false negative errors are greater than those listed in Table 3 [D.L. Lewis, 1990].

Standard deviations used in defining MDL or LOD and RDL values must reflect random measurement variability at or near these levels. Variability in repeated blank measurements is one recommended source for this information. Variability in results for repeated measurements of low-level standards (less than 10 times the MDL) is another source for this information. MDL or LOD estimates should be cautious at higher analyte concentrations since relative precision usually improves with increasing concentration. This may lead to a false impression that relative precision at the MDL or LOD is better than what can actually be achieved.

In addition to variability estimates used in defining MDL, LOD, and RDL from measurements in the proper region of the working range, it is also important to be aware of other sources of random variability in the measurement process. The data user must remember that the MDL, LOD, and RDL represent the capability of a given measurement system under a defined set of conditions. Typically the variability estimates used to develop the system capability estimates do not include sources of contamination or variation outside the laboratory (e.g. variability and contamination from sampling sources as described in Section I). Even when the laboratory result indicates the presence of an analyte (i.e., a signal >MDL or LOD) this does not guarantee that the detected analyte is representative of what the sample was intended to characterize. For example, if a bailer is rinsed with acetone prior to sampling groundwater and the water rinsate doesn't remove all of the acetone, then acetone may appear in the sample at the laboratory. In the absence of a field blank, these laboratory data would indicate that acetone was present in the groundwater samples because it was at a higher concentration than in the laboratory reagent blanks.

Several problems arise when applying the idealized model discussed above to real world measurement systems:

- The available knowledge about the random error components of the system is typically based on a limited amount of data, introducing considerable uncertainty with regard to the estimated standard deviation.
- Unavoidable contamination in trace analytical measurements often intro-

duces bias into the measurement system so that the average measured signal response from blanks is greater than zero.

- Low recoveries of the analytes of interest, due to calibration error or losses in the sample preparation process, may introduce additional bias error.
- Random error in many environmental measurements is not normally distributed, especially at very low concentrations.

The random error and recovery problems described above may be dealt with using rigorous statistical methods. However, such rigorous treatment is often inappropriate or unnecessary for the intended purpose, which is to derive estimates that aid in understanding the levels at which the data become meaningful. Since the objective is to aid data interpretation, rather than to make definitive probabilistic statements, it is sufficient for most environmental studies to recognize that these estimates are not exact; they are subject to certain limitations resulting from the acceptance of simplifying assumptions [D.L. Lewis, 1990].

A major use for RDLs is planning adequate measurements, whereas LODs are used for making decisions after measurements are complete. It must also be remembered that erroneous assumptions (e.g., using a reagent blank instead of a method blank) affect all results, not just low-level data [Currie, 1990].

RELATIVE BIAS AT LOW CONCENTRATION LEVELS

The second problem described above, bias, may be dealt with by accounting for bias in the calculated values of LOD and RDL. Background signal response due to low-level contamination should be taken into account by calculating the LOD and RDL as a given number of standard deviations above either within-batch blanks or the average blank concentration, rather than counting from zero concentration. This calculation effectively corrects for this "fixed" bias by offsetting the LOD and RDL values from zero by an amount equal to the average fixed bias. In addition to fixed bias, measurements may also exhibit relative bias, representing systematic errors proportional to concentration. Relative bias results in measured values which are low by some approximately constant ratio, due to losses in sample collection, systematic calibration errors, or other similar sources. Relative bias is an important consideration only when bias is large (>20%), well characterized, and unavoidable. Relative bias in LOD and RDL calculation may be accounted for by adjusting the calculated values for the average relative bias [D.L. Lewis, 1990].

For each method and analyte of interest, the MDL or LOD and RDL should be estimated with either within-batch blanks or averaged blank values and standard deviations. The estimate of MDL or LOD is calculated as the within-batch blank or the average blank value plus three times the standard deviation of the test results, for a 99+% level of confidence or 2.33 times the standard deviation for a 99% level of confidence. The estimated RDL is calculated as the within-batch or the average blank value plus six times the standard deviation, for a 99+% level of confidence or 4.66 times the standard deviation for a 99% level of confidence. If bias-corrected MDL, LOD, or RDL values are reported, this must be clearly noted. The bias corrected values are obtained by multiplying the uncorrected values by a bias correction factor, to correct for recovery. The bias correction factor is calculated as the theoretical concentration divided by the mean concentration for each analyte.

It is particularly important to consider relative bias in the context of the RDL. When we refer to the RDL, we are implicitly referring to a true concentration (as opposed to a measured concentration). Specifically, it is the concentration at which there is an "acceptably low risk" of making a false negative error and failing to detect an analyte which is present *at that concentration*. If an analyte signal is present but not reported (i.e., no signal greater than or equal to the MDL or LOD), it is possible to assert that the analyte is not there at a (true) concentration greater than the detection limit, with little risk of being incorrect in that assertion. Because detection limits are used in the context of *true concentrations*, when they really are estimated from measurement data, the estimates must be adjusted for systematic differences (i.e., relative bias) between measured concentrations and actual concentrations. For example, if the RDL of a method is 50 parts per billion by volume (ppbv), we can say that there is an "acceptably low" risk (<1%) of failing to detect the analyte of interest. If the method has a negative bias of 40%, however, repeated measurements of samples for which the true concentration is 50 ppbv will result in measured values which average 30 ppbv (60% of 50 ppbv). However, since 30 ppbv represents 6σ, individual measurements may range from 15 to 45 ppbv (the values at 3σ and 9σ, that is, $6\sigma \pm 3\sigma$). If observed variability in repeated measurements by this method were used to calculate the RDL, without correcting for the negative bias, the calculated value would be 30 ppbv, corresponding to the uncorrected RDL estimate instead of the true RDL estimate of 50 ppbv [D.L. Lewis, 1990].

Correction for relative bias is less useful when applied to the MDL or LOD since these levels represent observed measurement "signals" or instrument response as the criterion for making a detection decision. Measurement signals below the MDL or LOD result in a decision of "not detected"; signals above the LOD result in a decision of "detected". Obviously, if no signal is observed, the result is automatically classified as "not detected". In making the detection decision, it is appropriate to use the uncorrected MDL or LOD as the decision criterion. The bias-corrected estimate is useful only to relate this observed response to a true concentration. For example, the measurement system described above, with a negative bias of 40% and a RDL of 50 ppbv, would have a LOD (uncorrected) of 15 ppbv (3σ lower than the uncorrected RDL of 30 ppbv, assuming an average background concentration of zero). A measured value of 14 ppbv would be classified as "not detected". A measured value of 15 ppbv would be classified as "detected", and would correspond to a true concentration of 25 ppbv (3σ lower than the RDL of 50 ppbv) corrected for relative bias [D.L. Lewis, 1990].

When is an Analyte Not Detected?

As discussed above, the MDL or LOD represents the level at which it becomes possible to make detection decisions. When we measure an analyte at or above this level, we can be confident that the analyte is in fact present at a concentration greater than would be expected for a blank. Furthermore, the RDL represents the likely analyte detection level when it is set at twice the MDL or LOD instead of equal to it. In other words, there is very little chance of failing to detect an analyte present at a concentration equal to the (bias-corrected) RDL concentration. The RDL is therefore useful as a guide for designing environmental analyses. If an analyte is reported to be "not reliably detected", it is safe to conclude that it is not there at a concentration at or above the reliable detection limit. The "detected"/"not detected" decision, however, still is based on the MDL or LOD and whether or not a signal greater than this level is observed.

The ASTM *Standard Practice for Intralaboratory Quality Control Procedures and a Discussion on Reporting Low-Level Data* states that, "in general, the analytical chemist will rarely have responsibility for inference from data sets or even be in a position to know which data may be combined. Therefore censoring of results to prevent a possible faulty inference being drawn from an individual datum represents an unwarranted assumption of responsibility; yet the value judgment of the chemist on the data should be known or expressed relative to quality control information developed. Results reported as 'less than' or 'below the criterion of detection,' are virtually useless for estimating outfall and tributary loadings or concentrations, for example."

The above statement is directly opposed to the earlier ACS recommendations [Keith, et al., 1983], yet even within current ASTM committees involved with measuring and reporting environmental analytical data, there is significant opposition to the reporting of values below the MDL or LOD. The cause of this schizophrenic attitude in reporting data below the MDL or LOD is based largely on past observations and experiences of such data without reference to its confidence limits in other summary reports, databases, and other calculations.

Some EPA reports have recommended the opposite view—that when values are reported as being less than a MDL, they should be included as values *at* the MDL. This approach would, of course, produce highly biased data and is as bad a practice (many would say worse) as discarding data.

REPORTING BY USERS

Who, then should determine when and how to report data below a selected value? If, as is often true with many analytical laboratories, the analyst is not in a position to know how the data may ultimately be used, who should decide when and how to report low-level data? The decision as to when to accept that an analyte is present in a sample is relatively straightforward; it is virtually unanimously accepted that an analyte should be reported as present when it is measured at a concentration 3σ or more above the average corresponding method blank. The significance of these results from a data user standpoint depends on the concentration of the analytes of interest found in the field blanks. The larger problem, as we have seen from the preceding discussion, is deciding when an analyte is not present in a sample. Basically, the user of the analytical data is most often in the best position to make this decision. Beyond that, the user is responsible for this decision—not the laboratory. The laboratory should report all the data to the user for the user to interpret.

Many environmental analytical laboratories today subscribe to the practice of not reporting data less than the detection limit (MDL or LOD). Many EPA methods require reporting MDLs, but the user of this data should be aware that MDLs have limitations in their usefulness for data interpretation. These limitations arise from the manner in which the MDL is determined as specified in U.S. government publications. The MDL often is less than or equal to the LOD. One of the limitations of using MDLs is that calculations involving them do not include any persistent background contributions. For example, common laboratory solvents are often found as contaminants in blanks, at levels above the MDL.

The MDL allows distinction between a real analyte signal and true zero concentration; however, MDL does not allow distinction between the analyte signal due to contamination (in a blank) and the analyte signal from the sample. Therefore, the MDL does not allow assessment of the true significance of the sample results. In other words, MDL does not provide a criterion for judging whether the sample results are significantly different from the blanks, because the MDL calculation considers only the variability in the measurement process, not the bias due to consistent blank contamination. Conversely, the LOD is defined as a value a certain selected number of standard deviations above the average background concentration (i.e., above the average concentration found in the method blanks).

In the case where there is no contamination (i.e., the average background value is zero), then the MDL and LOD are the same.

MDLs are most useful for relative comparisons of:

- Performance between different laboratories
- Performance between different matrices
- Performance between different methods
- Performance over time

LIMIT OF QUANTITATION (LOQ)

The *limit of quantitation* (LOQ) is defined as the level above which quantitative results may be obtained with a specified degree of confidence. LOQ is different, and more difficult, than measuring the presence or absence of an analyte. Confidence in the apparent analyte concentration increases as the analyte signal increases above the LOD. The value for LOQ $= 10\sigma$ is recommended, corresponding to an uncertainty of $\pm$ 30% in the measured value ($10\sigma \pm 3\sigma$) at the 99% confidence level [Keith, et al., 1983]. In other words, when the analyte signal is 10 or more times larger than the standard deviation of the measurements there is a 99% probability that the true concentration of the analyte is $\pm$ 30% of the calculated concentration. This assumes a well characterized blank has provided background signal response.

The LOQ is most useful for defining the lower limit of the useful range of concentration measurement methodology. The useful range extends from this lower value to an upper value, where the response is no longer linear and sometimes is referred to as the limit of linearity (LOL).

The LOD, RDL, and LOQ are shown graphically in Figure 12. The base scale is in *standard deviations* of the measurement process, which is assumed to be the same for all of the measurements indicated.

The environmental matrix can significantly affect the relative level at which quantitation can reliably occur, especially when it is expressed in terms of the MDL, which neglects background effects. A *Practical Quantitation Limit* (PQL) has been defined by the EPA as the lowest level that can be reliably achieved within specified limits of precision and accuracy during routine operating conditions. The

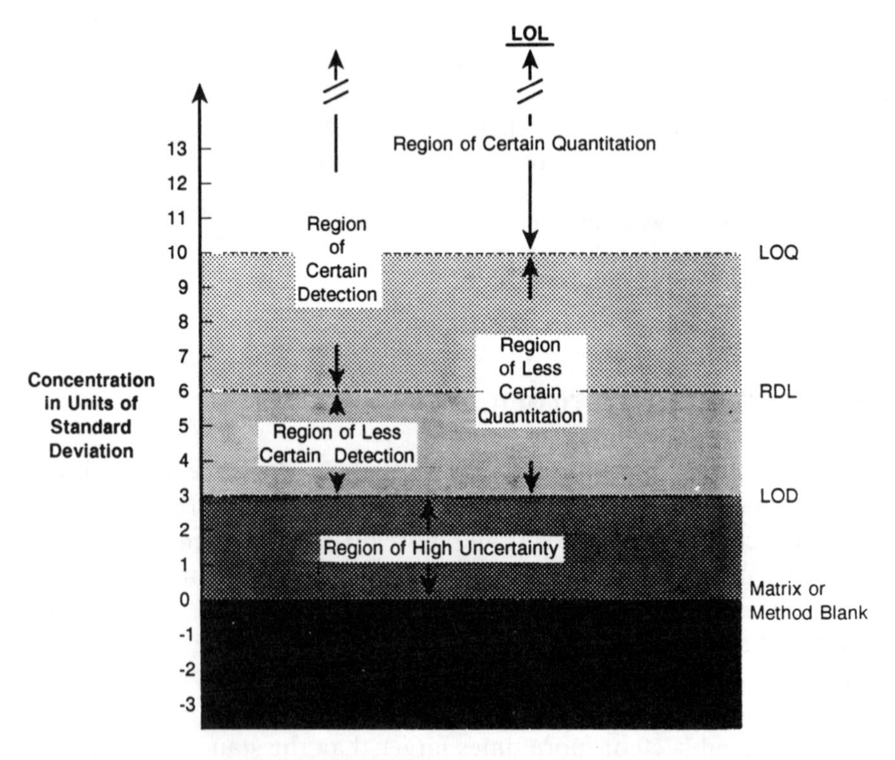

Figure 12. The relationship of LOD, RDL, and LOQ to signal strength (analyte concentration). The recommended LOD is 3σ, the recommended RDL is 6σ, and the recommended LOQ is 10σ.

Table 4. Guidelines for Laboratories Reporting Data

Analyte Concentration in Units of σ ($S_t - S_b$) for LOD or Just S_t for MDL	Regions of Relative Reliability	Report As
"Zero"	No observed signal	ND (MDL or LOD = X)
$<3\sigma$	Region or high uncertainty (<MDL or <LOD)	Y* (MDL or LOD = X)
$>3\sigma$	Method detection limit (MDL) or limit of detection (LOD) and above	Y (MDL or LOD = X)

"Zero" may be a negative value or no discernable detector response.
S_t = Total value measured for the test sample analyte.
S_b = Corresponding value for the blank.
Y = Reported sample concentration.
X = Calculated MDL or LOD.
* = Data below the MDL or LOD should be flagged. The flag notation varies with method requirements and is not important, but the key to all notations must always be included.

PQL may be 10 to 100,000 times or more greater than the MDL for a method.

An excellent brief review of common statistical techniques used for evaluating analytical data is presented in Chapter 4 of John Taylor's book, *Quality Assurance of Chemical Measurements* [1987]. Subjects covered include data distribution, measurement statistics, standard deviation estimation (σ), confidence limits, confidence intervals, statistical tolerance intervals, outliers, and control charts. Examples of calculations are provided along with the necessary equations.

GUIDELINES FOR REPORTING AND PRESENTING DATA

Laboratories report data they generate from analyses to users. Data users interpret these data and present them with discussion and/or interpretation in documents, reports, summaries, etc.

Revised guidelines for reporting data by laboratories (based on the above distinctions) are given in Table 4. In using the table, remember that the concentration levels indicated refer to interpretation of single measurements. Users typically work with data sets composed of several or many such individual measurements.

Since it is not a laboratory responsibility to interpret or censor data, all data should be reported, but these data also must be completely documented with respect to problems and limitations, especially data near or below the MDL or LOQ.

If the data user determines that the data may potentially be taken

Table 5. Guidelines for Users Presenting Data[a]

Liberal Guidelines		Conservative Guidelines	
Measured Value	**Measured Report As**	**Value**	**Report As**
"Zero"	ND (RDL = X)	"Zero"	ND (RDL = X)
<LOD or <MDL	Y* (MDL or LOD = X)	<LOD or <MDL	Less than LOD OR MDL (LOD or MDL = X)
>LOD or >MDL but <LOQ	Y** (LOQ = Z)	>LOD or >MDL but <LOQ	Y** (LOQ = Z)
>LOQ	Y	>LOQ	Y

[a]"Zero" may be a negative value or no discernable detector response. X and Y are specific values calculated and measured, respectively. The same guidelines for flagging data <LOQ apply as in Table 4. Generally, different flags should be used when the data is below MDL or LOD and LOQ. In this example * designates a flag for data below the MDL or LOD and ** designates a flag for data below the LOQ.

and used beyond the limits defining its measurement reliability, a conservative approach should be taken, the signals below 3σ should be presented as "less than LOD" or "less than MDL", and the corresponding LOD or MDL should be given. If no signal above background is detected then "not detected" or ND should be presented with the RDL to distinguish this situation from that where a small signal below the LOD is distinguishable from the background. Referencing LOD or MDL with ND for the liberal presentation format and the RDL with ND for the conservative presentation format is another acceptable approach.

If the data user determines that the data potentially is being used improperly, a liberal approach should be taken in which all values be presented (except outliers which are discussed later) along with the MDL or LOD. Signals in the "region of less-certain quantitation" (3σ to 10σ) should be presented with their concentrations, and the LOQ should be next to them either in an adjacent column or in parenthesis. The guidelines for these report format presentation decisions are listed in Table 5.

An example of these two user presentation formats is shown below in Table 6 where:

Table 6. Low-Level Data Formats

Sample	Liberal	Conservative
A	ND (RDL = 65 μg/L)	ND (RDL – 65 μg/L)
B	25 μg/L* (LOD = 35 μg/L)	Less than LOD (LOD = 35 μg/L)
C	50 μg/L** (LOQ = 105 μg/L)	50 μg/L** (LOQ = 105 μg/L)
D	110 μg/L	110 μg/L

- The average well characterize blank value = 5 μ/L
- The standard deviation (σ) of the blank = 10 μg/L
- The LOD of 3σ above the blank = 35 μg/L (3 X 10 + 5)
- The RDL of 6σ above the blank = 65 μg/L (6 X 10 + 5)
- The LOQ of 10σ above the blank = 105 μg/L (10 X 10 + 5)
- Measured analyte values are "zero", 25, 50, and 110 μg/L in four samples labeled A, B, C, and D respectively

Notice that the words "less than" are recommended instead of the other commonly used format "<LOD" or "<MDL". This is because tables of data are frequently passed electronically from one report to another or one database to another. During these transmissions, if different word processors or database programs are used, it is easy for the "less than" and "greater than" characters to be stripped out and lost. The result is that the LOD or MDL number appears by itself and is interpreted as a higher "true" value. A similar problem will be presented if ND is transferred without the corresponding LOD or RDL value.

The symbols "T" or "tr" for amounts and the term "trace" and similar statements of relative concentration should be avoided because of the relative nature of such terminology, the confusion surrounding it, and the danger of its misuse [Keith, et al., 1983].

Data measured at or near the LOD have two problems. First, the uncertainty of its value can approach, and even equal, the reported value. Figure 13 illustrates this theoretical concept in relation to the LOD and LOQ using units of standard deviation (σ). Second, confirmation of the species reported is virtually impossible unless the identification uses highly selective methods (such as mass spectrometry). These problems diminish when reliably measurable amounts of analytes are present. Accordingly, legal and regulatory actions usually should be limited to data at or above the reliable detection level, or they should be conservatively above the LOQ. Exceptions to this logical guideline may be expected when the chemical analytes involved are so hazardous to human health that realistic safety factors push action levels below their reliable measurement levels. Relatively few pollutants are known to be so toxic or hazardous that they meet this reasonable criterion.

It must be emphasized that the MDL, LOD, RDL, and LOQ are not intrinsic constraints of the analytical methodology. They depend upon the precision attainable by a specific laboratory, working with a specific matrix, when using that methodology. Thus, MDLs, LODs, RDLs, and LOQs can be very diverse. Unfortunately, this basic fact generally is not considered when evaluating environmental

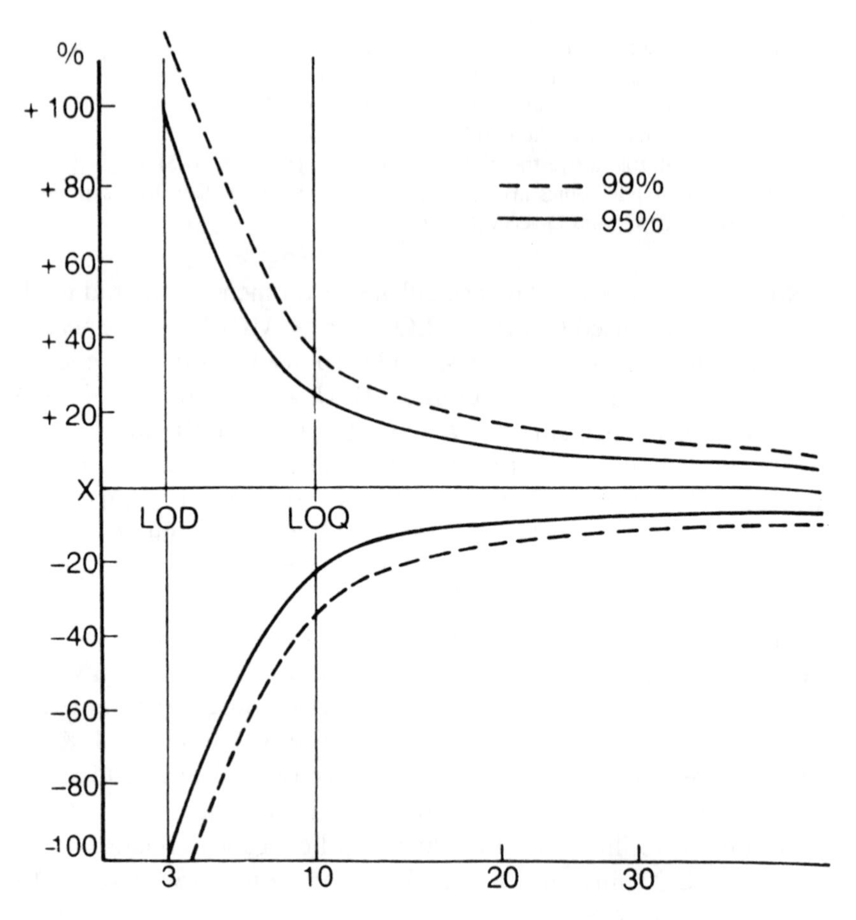

Figure 13. Uncertainty of measurement close to the limit of detection. Reprinted from J.K. Taylor, 1987, p. 82, with permission from Lewis Publishers.

analytical data. Most users of analytical data are unaware of this caveat. Published values of MDLs must be considered only typical. Each laboratory reporting data must evaluate its own precision and estimate its own MDL, LOD, RDL, and LOQ values for analytes of interest, for each type of matrix it analyzes. A common and acceptable alternative, when method-specified limits are available (for example with many EPA methods), is to verify that each instrument can meet or exceed these published limits. If a method has any possible sensitivity to operator variability, the instrument and method verification should be performed by each person who will use it. *Method sensitivity* in this context is defined as the rate of change in instrument response to the change in analyte concentration (i.e., the slope of the calibration curve); it should not be confused with the

LOD. Such method verification is expensive, but without it, quantitative and sometimes qualitative data can have no confidence placed in them.

There are also upper levels of reliable measurements. These vary from method to method and are a function of a particular instrument's detector response to each specific analyte. At high concentration levels (a term that is relative to each analyte and detector considered) measurements will become nonlinear with increasing concentration. This is called the *limit of linearity* (LOL) and is usually caused by the analyte chemically or physically saturating the detector. Figure 14 illustrates the useful range of a typical method of measurement.

The analytical chemist is responsible for fully describing and interpreting the data and reporting them in an appropriate manner. Remember that all users of those data will depend (perhaps years later) on how clearly and thoroughly the data were recorded and described.

Measurement results should be expressed so that their meaning is not distorted by the reporting process. The public at large will not be able to recognize that 10,000 ng/kg and 10 μg/kg are the same. In

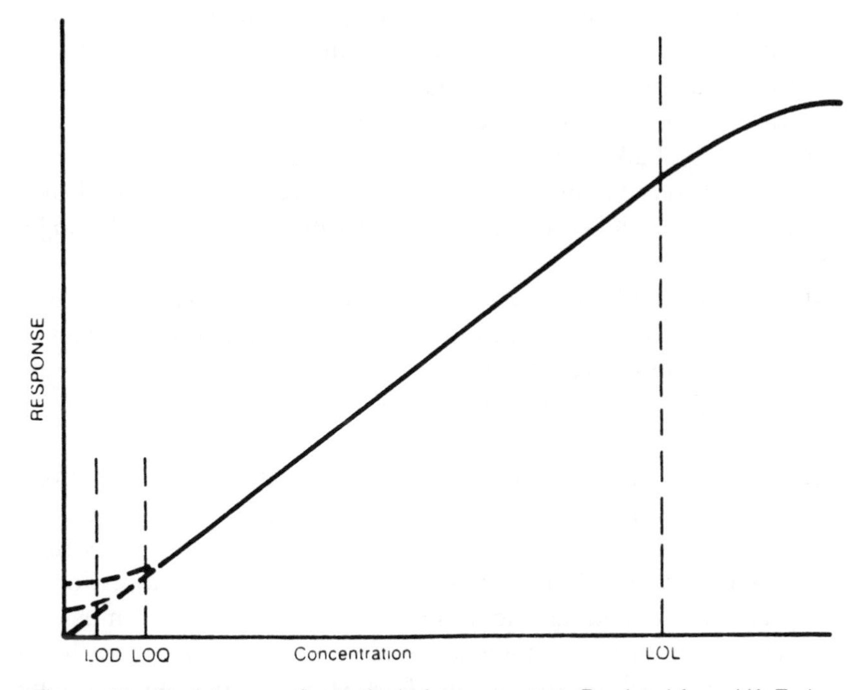

Figure 14. Useful range of a method of measurement. Reprinted from J.K. Taylor, 1987, p. 80, with permission from Lewis Publishers.

general, μg/kg (parts per billion) are commonly employed with most ambient environmental samples; ng/kg (parts per trillion) are sometimes employed with very low level analytes in potable (drinking) water and human samples. Since parts per million, billion, trillion, etc., are less definitive than specific units of measurement such as mg/L, $\mu g/cm^3$, ng/kg, etc., the use of these more specific units for expression of concentration should always be used.

Generally, analytical values from a laboratory should be reported as measured (uncorrected for recovery) with full and complete supporting data involving recovery experiments. If the measurements are reported as "recovery-corrected", all calculations and experimental data should be documented so that the original uncorrected values can be derived if desired. In carrying out recovery studies, the analyst should recognize that an analyte added to a blank sample may behave differently (typically, showing higher recovery) than that analyte in a test sample. In such cases, the method of standard addition tends to lead to erroneously low values [Keith, et al., 1983].

Laboratory reports must contain sufficient data and information so that users of the conclusions (even years later) can understand the interpretations from raw data, without having to make their own. Unless this objective is achieved, the samplers and analysts have not done their jobs properly, and the users of the data have every right to complain that their objectives and goals were not completely met. Laboratory reports also must clarify which results, if any, have been corrected for blank and recovery measurements. Generally corrections for recovery are not made, but percent recovery should always be reported where it is involved. Any other limitations should also be noted.

If published methodology is used, it must be cited. Any modifications, as well as any new methodology techniques, new approaches in making test sample measurements, or interpretations of results, must be described in detail, including test results and details of their validation.

Raw data for each sample, along with data from reagent blanks, control, spiked samples, and all other quality control samples, should be suitably identified if included in laboratory reports (and QC data usually are). If average values are reported, an expression of the precision, including the number of measurements, must be included. Details of the analytical results should be written with the standard deviation and the mean. They should show the averaging process accounts for sample heterogeneity as well as observed imprecision among replicate measurements of homogenized samples [ACS].

GUIDELINES FOR HANDLING DATA

Electronic data handling, reduction, storage, and transmission systems greatly facilitate data handling and help minimize errors due to misreading, faulty transcription, or miscalculation. However, the performance of the data system should be tested periodically with known data that have already been calculated; this testing should be part of the quality assurance program. These tests must have sufficient diversity and rigor to provide a reliable test of the data handling system [Keith, et al., 1983]. Be especially alert for the deletion of certain characters such as > or <, *, "flags", or other notes of explanation when data are passed around electronically.

In performing mathematical operations or calculations, the preferred protocol is for measured or observed data to be recorded with as many numbers as possible; rounding numbers should be deferred until all calculations have been made and their statistical significance has been evaluated. The number of significant figures is the number of digits remaining when the data are so rounded. The last digit, or at most the last two digits, are expected to be the only ones that would be subject to change on further experimentation. Thus, for a measured value of 21.5, only the 5, and at most the 1.5, would be expected to be subject to change. Such data would be described as having three significant figures. In counting significant figures, any zeros used to locate a decimal point are not considered as significant. Thus 0.0025 contains only two significant figures. Any zeros to the right of the digits are considered significant; thus, 2500 and 2501 each have four significant figures. Only those that have significance should be retained. Zeros should not be added to the right of significant digits to define the magnitude of a value unless they are significant, since this would confuse the significance of the value. For example, it is not good practice to report a value as 2500 ng, but rather 2.5 μg if the data are reliable to two significant figures. The use of exponential notation, e.g., 3.5×10^3 is an acceptable way to express both the number of significant figures and the magnitude of a result [Taylor, 1987].

If possible, and within the scope of desired results, a number of measurements sufficient for statistical treatment should be made. Three measurements, as a minimum, are required to calculate standard deviations. When no statistical treatment is made, an explanation is necessary, including complete details of the treatment of the data.

Since laboratories generate data for their clients (users), laboratories are not the final step in the data use process; therefore, rounding performed by the laboratory should attempt to preserve measurement variability. A good rule is for the laboratory to retain at least one significant figure beyond that known with reasonable certainty. Also, the laboratory should not attempt to convey measurement uncertainty through use of significant figures—this information should be provided in accompanying statements of precision and accuracy. The data user provides the final step in presenting and working with data sets. This is the point at which rounding of data should occur [D.L. Lewis, 1990].

The following rules for rounding data, consistent with its significance, are summarized in Chapter 22 of Taylor's book [1987]:

- When the digit immediately after the one to be retained is less than five, the retained figure is kept unchanged. For example: 2.541 becomes 2.5 to two significant figures.
- When the digit immediately after the one to be retained is greater than five, the retained figure is increased by one. For example: 2.453 becomes 2.5 to two significant figures.
- When the digit immediately after the one to be retained is exactly five and the retained digit is even, it is left unchanged but when it is exactly five and the retained digit is odd the retained digit is increased by one. For example: 3.450 becomes 3.4, but 3.550 becomes 3.6 to two significant figures.
- When two or more figures are to the right of the last figure to be retained, they are considered as a group in rounding decision. Thus in 2.4(501), the group (501) is considered to be greater than 5, while for 2.5(499), (499) is considered to be less than 5.

FINAL COMMENTS

Some of the above is common sense, but large portions of these practical recommendations are tried and true procedures from experienced environmental analytical chemists with many years of experience. There are many pitfalls with environmental analyses because of their complexity, their extremely low detection and quantitation levels, and the usual separation of the sampling protocols and people from the analytical protocols and people. Unfortunately much of the available environmental analytical data is either of uncertain or less certain confidence than the users and purchasers of it believe—even when it has been collected and analyzed according to published protocols. Hopefully, this ACS guide will help planners, users, and analysts more critically plan, measure, and evaluate environmental analytical data. But remember that even mediocre data often can be useful as long as it is recognized as such. If you recognize these differences—and change them if time, money, and desires permit it—many of your analytical objectives may be achievable, and at least your data will be of a known quality, as good or as bad as that may be.

REFERENCES

ACS Committee on Environmental Improvement. "Guidelines for Data Acquisition and Data Quality Evaluation in Environmental Chemistry," *Anal. Chem.*, 52:2242–2248 (1980).

ASTM. "Standard Practice for Intralaboratory Quality Control Procedures and a Discussion on Reporting Low-Level Data," ASTM D4210–83, Philadelphia, PA, 1983.

Alabert, R. and W. Horwitz. "Coping With Sampling Variability in Biota: Percentiles and Other Strategies," in *Principles of Environmental Sampling*, L.H. Keith, Ed. (Washington, D.C.: American Chemical Society, 1988), p. 337.

Barcelona, M.J. "Overview of the Sampling Process," in *Principles of Environmental Sampling*, L.H. Keith, Ed. (Washington, D.C.: American Chemical Society, 1988), p. 3.

Barcelona, M.J. personal communication (July, 1989).

Barnett, V. and T. Lewis. "Outliers in Statistical Data," (Bath, England: John Wiley & Sons, 1978).

Blacker, S.M. "A New QC Alternative," *Environmental Lab*, 22, April/May 1990.

Black, S.C. "Defining Control Sites and Blank Sample Needs," in *Principles of Environmental Sampling*, L.H. Keith, Ed. (Washington, D.C.: American Chemical Society, 1988), p. 109.

Bone, L.T. "Preservation Techniques for Samples of Solids, Sludges and Nonaqueous Liquids," in *Principles of Environmental Sampling*, L.H. Keith, Ed., (Washington, D.C.: American Chemical Society, 1988), p. 409.

Borgman, L.E. and W.F. Quinby. "Sampling for Tests of Hypothesis When Data are Correlated in Space and Time," in *Principles of Environmental Sampling*, L.H. Keith, Ed., (Washington, D.C.: American Chemical Society, 1988), p. 25.

Burns, M. personal communication (July, 1989).

Cheeseman, R.V. and A.L. Wilson. "Manual on Analytical Quality Control for the Water Industry," Technical Report TR66, (England: Water Research Center, 1978), pp. 38–41.

Clements, J.B. personal communication (July, 1989).

Clements, J.B. and R.G. Lewis. "Sampling for Organic Compounds," in *Principles of Environmental Sampling*, L.H. Keith, Ed., (Washington, D.C.: American Chemical Society, 1988), p. 287.

Coakley, W. personal communication (July, 1989).

Cowgill, U.M. "Sampling Waters: The Impact of Sample Variability on Planning and Confidence Levels," in *Principles of Environmental Sampling*, L.H. Keith, Ed. (Washington, D.C.: American Chemical Society, 1988), p. 171.

Cowgill, U.M. personal communication (July, 1989).

Cox, G.V. personal communication (July, 1989).

Currie, L.A. "Limits for Qualitative Detection and Quantitative Determination," *Anal. Chem.*, 40:586–593, (1968).

Currie, L.A. "Detection: Overview of Historical, Societal and Technical Issues," in *Detection in Analytical Chemistry, Importance, Theory and Practice*, L.A. Currie, Ed. (Washington, D.C.: ACS Symposium Series No. 361, American Chemical Society, 1988).

Currie, L.A. personal communication (May, 1990).

Donaldson, W.T. personal communication (July, 1989).

Dupuy, A.E. personal communication (July, 1989).

Dupuy, A.E. personal communications (April, 1990).

Englund, E.J. personal communication (July, 1989).

Ferrario, J.B. personal communication (July, 1989).

Flatman, G.T. personal communication (July, 1989).

Forsberg, K. "Chemical Protective Clothing Performance Index," L.H. Keith, ed., Instant Reference Sources, Inc., Austin, TX, 1990.

Garner, F.C., M.A. Stapanian, and L.R. Williams. "Composite Sampling for Environmental Sampling," in *Principles of Environmental Sampling*, L.H. Keith, Ed. (Washington, D.C.: American Chemical Society, 1988), p. 363.

Gilbert, R.O. "Statistical Methods for Environmental Pollution Monitoring," (New York: Van Nostrand Reinhold, 1987).

Haeberer, F. personal communication (May, 1989).

Haeberer, F. personal communication (May, 1990).

Haile, D. personal communication (July, 1989).

Hicks, B.B., T.P. Meyers, and D.D. Baldocchi. "Aerometric Measurement Requirements for Quantifying Dry Deposition," in *Principles of Environmental Sampling*, L.H. Keith, Ed. (Washington, D.C.: American Chemical Society, 1988), p. 297.

Ho, J.S.-Y. "Effects of Sampling Variables on Recovery of Volatile Organisms in Water," *J. AWWA*, 583–586 (1983).

Homsher, M.T. personal communication (July, 1989).

Horowtiz, W. personal communication (August, 1989).

Hunt, D.T.E. and A.L. Wilson. "The Chemical Analysis of Water," The Royal Society of Chemistry, England, 1986.

Journel, A.G. personal communication (July, 1989).

Journel, A.G. and F. Alabert. *Terra Nova*, 1,2:123–134 (1989).

Keith, L.H. in *Dynamics, Exposure and Hazard Assessment of Toxic Chemicals*, R. Haque, Ed. (Woburn, MA: Ann Arbor Science Publishers, 1980), pp. 41–45.

Keith, L.H., W. Crummett, J. Deegan, Jr., R.A. Libby, J.K. Taylor, and G. Wentler. "Principles of Environmental Analysis," *Anal. Chem.* 55:2210–2218 (1983).

Keith, L.H. "Defining QA and QC Sampling Requirements. Expert Systems as Aids," in *Principles of Environmental Sampling*, L.H. Keith, Ed. (Washington, D.C.: American Chemical Society, 1988).

Keith, L.H. *Principles of Environmental Analyses*, (Washington, D.C.: American Chemical Society, 1988).

Keith, L.H. "Environmental Sampling: A Summary," *ES&T 24* 5:610–617 (1990).

Keith, L.H., Ed. *Sampling and Analysis Methods Database,* Vols. I–III, (Chelsea, MI: Lewis Publishers, 1990).

Kent, R.T. and K.E. Payne. "Sampling Groundwater Monitoring Wells: Special Quality Assurance and Quality Control Procedures," in *Principles of Environmental*

Sampling, L.H. Keith, Ed., (Washington, D.C.: American Chemical Society, 1988), p. 231.

Kirchmer, C.J. "Estimation of Detection Limits for Environmental Analytical Procedures," in *Detection in Analytical Chemistry, Importance, Theory and Practice*, L.A. Currie, Ed. (Washington, D.C.: ACS Symposium Series 361, American Chemical Society, 1988).

Kirchmer, C.J. personal communication (April, 1990).

Kulkarni, S. personal communication (August, 1989).

Lewis, D.L. "Assessing and Controlling Sample Contamination," in *Principles of Environmental Sampling*, L.H. Keith, Ed. (Washington, D.C.: American Chemical Society, 1988), p. 119.

Lewis, D.L. personal communication (January 1990).

Lewis, R.G. "Problems Associated With Sampling for Semivolatile Organic Chemicals in Air," *Proceedings, 1986 EPA/APCA Symposium on Measurement of Toxic Air Pollutants*," (Pittsburgh, PA: APCA Special Publication VIP-7 Air and Waste Management Association, 1986), pp. 134–145.

Lewis, R.G. personal communication (July, 1989).

Liggett, W.S. "Assessment of Measurement Uncertainty: Designs for Two Heteroscedastic Error Components," in *Principles of Environmental Sampling*, L.H. Keith, Ed. (Washington, D.C.: American Chemical Society, 1988), p. 191.

Lodge, J.P. personal communication (July, 1989).

Maskarinec, M.P., C.K. Bayne, L.H. Johnson, S.K. Holladay, and R.A. Jenkins. "Stability of Volatile Organics in Environmental Water Samples: Storage and Preservation," Oak Ridge National Laboratory Report ORNL/TM-11300, ORNL, TN, August 1989; and "Stability of Volatile Organic Compounds in Environmental Water Samples During Transport and Storage," *ES&T*, 24 1665–1670.

McCormack, M. "Realistic Detection Limits and the Political World," in *Detection in Analytical Chemistry, Importance, Theory and Practice*, L.A. Currie, Ed. (Washington, D.C.: ACS Symposium Series No. 361, American Chemical Society, 1988).

McKinney, J.D. et al. in *Environmental Health Chemistry*, McKinney, J.D., Ed.; (Woburn, MA: Ann Arbor Science Publishers, 1980), p 441.

Meier, E.G. personal communication (July, 1989).

Messner, M.J. personal communication (November, 1989).

Newburn, L.H. "Modern Sampling Equipment: Design and Application," in *Principles of Environmental Sampling*, L.H. Keith, Ed., (Washington, D.C.: American Chemical Society, 1988), p. 209.

Newman, M.C., P.M. Dixon, B.B. Looney, and J.E. Pinker, III. "Estimating Mean and Variance for Environmental Samples with Below Detection Limit Observations," *Water Resources Bulletin*, 25, 4:95 (1989).

Norris, J.E. "Techniques for Sampling Surface and Industrial Waters: Special Considerations and Choices," in *Principles of Environmental Sampling*, L.H. Keith, Ed., (Washington, D.C.: American Chemical Society, 1988), p. 247.

Parr, J.L., M. Ballinger, O. Callaway, and K. Carlberg. "Preservation Techniques for Organic and Inorganic Compounds in Water Samples," in *Principles of Environmental Sampling*, L.H. Keith, Ed. (Washington, D.C.: American Chemical Society, 1988), p. 221.

Parr, J.L., K. Carlberg, and M. Bollinger personal communication (September, 1989).

Piwoni, M.D. personal communication (July, 1989).

Rogers, L.B. et al. "Recommendations for Improving the Reliability and Acceptability of Analytical Chemical Data Used for Public Purposes," *Chem. Eng. News*, 60,23:44 (1982).

Smith, J.S., D.P. Steele, M.J. Malley, and M.A. Bryant. "Groundwater Sampling," in *Principles of Environmental Sampling*, L.H. Keith, Ed., (Washington, D.C.: American Chemical Society, 1988), p. 225.

Spittler, T.D. and J.B. Bourke. "Considerations for Preserving Biological Samples," in *Principles of Environmental Sampling*, L.H. Keith, Ed., (Washington, D.C.: American Chemical Society, 1988) p. 225.

Stapanian, M.A. and F.C. Garner. personal communication (July, 1989).

Steele, D.P. personal communication (July, 1989).

Tanner, R.L. "Airborne Sampling and In Situ Measurement of Atmospheric Chemical Species," in *Principles of Environmental Sampling*, L.H. Keith, Ed., (Washington, D.C.: American Chemical Society, 1988), p. 275.

Taylor, J.K. "What is Quality Assurance," Proceedings of the Conference on Quality Assurance of Environmental Measurements, Boulder, CO, ASTM, Philadelphia, PA, (August 1983).

Taylor, J.K. "Validation of Analytical Methods," *Anal. Chem.*, 55:600A-608A (1983).

Taylor, J.K. *Quality Assurance of Chemical Measurements*, (Chelsea, MI: Lewis Publishers, Inc., 1987.

Taylor, J.K. "Defining the Accuracy, Precision and Confidence Limits of Sample Data," in *Principles of Environmental Sampling*, L.H. Keith, Ed. (Washington, D.C.: American Chemical Society, 1988), p. 101.

Trehey, M.L. and T.I. Bieber. "Detection, Identification and Quantitative Analysis of Dihaloacetonitriles in Chlorinated Waters," in *Advances in the Identification and Analysis of Organic Pollutants in Water*, L.H. Keith, Ed. (Ann Arbor, MI: Ann Arbor Science Publishers, Inc., 1981), p. 491.

Triegal, E.K. "Sampling Variability in Soils and Solid Wastes," in *Principles of Environmental Sampling*, L.H. Keith, Ed. (Washington, D.C.: American Chemical Society, 1988), p. 385.

U.S. EPA, *Methods for Evaluating the Attainment of Cleanup Standards. Volume 1: Soils and Solid Media*, Washington, D.C., EPA 320/02-89-042, February 1989.

Walker, M.M. personel communication (July, 1989).

Ziman, S.D. personal communication (July, 1989).

OTHER USEFUL SOURCES OF INFORMATION

ASTM, D3370-76, "Standard Practices for Sampling Water," *ASTM Annual Book of Standards, Part 31*, (Philadelphia: American Society for Testing and Materials.)

Barcelona, M.J., J.P. Gibb, J.A. Helfrich, and E.D. Garski. "Practical Guide for Groundwater Sampling," EPA 600/S2-85/104, Robert S. Kerr Envr. Research Laboratory, Ada, OK (1985).

Collins, A.G. and A.I. Johnson, Eds. "Groundwater Contamination Field Methods," *ASTM Special Technical Publication 963*, Philadelphia, 1988.

Currie, L.A. "Detection in Analytical Chemistry, Importance, Theory and Practice," *ACS Symposium Series No. 361*, (Washington, D.C.: American Chemical Society, 1988).

Currie, L.A. "The Limitations of Models and Measurements as Revealed Through Chemometric Intercomparison," *J. Research of the NBS*, 90,6: Nov.-Dec. (1985).

Currie, L.A. "Scientific Uncertainty and Societal Decisions: The Challenge to the Analytical Chemist," *Anal. Letters*, 13,A1:1-31 (1980).

Currie, L.A. in "Treatise on Analytical Chemistry," I.M. Kolthoff and P.J. Elving, Eds., 2nd Edition, Part 1, Vol 1, (London: Wiley, 1978), pp. 95-242.

DeVera, E.R. et al. "Samplers and Sampling Procedures for Hazardous Waste Streams," EPA 600/2-80-018, January 1980.

Driscoll, F.G. "Groundwater and Wells," Second ed., Johnson Division, St. Paul, MN, 1986.

Driscoll, F.G. "Groundwater and Wells," Second ed., Johnson Division, St. Paul, Minn. 55112, 1987.

Freedman, D. *Symposium on Waste Testing and Quality Assurance, Vol. 3*, (Philadelphia: ASTM Special Technical Publication 1075, 1989).

Glaser, J.A., D.L. Forest, G.D. McKee, S.A. Quave, and W.L. Budde. "Trace Analysis for Wastewaters," *ES&T*, 15:1426 (1981).

Journel, A.G. and I. Huijbregts. "Mining Geostatistics," (New York: Academic Press, 1981).

Journel, A.G. "Geostatistics for the Environmental Sciences," U.S. EPA Project Report CR 811893, Washington, D.C., 1987.

Kratochvil, B. and J.K. Taylor "Sampling for Chemical Analysis," *Anal. Chem.*, 53,8 (1981).

Lodge, J.P. "Methods of Air Sampling and Analysis," (Chelsea, MI: Lewis Publishers, Inc., 1989).

Mason, B.J. "Preparation of Soil Sampling Protocols: Techniques and Strategies," Report EPA/600/4-83/020, EMSL, Las Vegas, NV, March 1982.

Perket, C. *Quality Control in Remedial Site Investigation, Vol. 5*, Philadelphia: ASTM Special Technical Publication 925, 1986).

The Conservation Foundation. *Sources of Contamination, in Groundwater Protection*, Washington, D.C., 1987.

U.S.D.A. *Chemistry Quality Assurance Handbook*, (Washington, D.C.: U.S. Department of Agriculture, Food Safety and Inspection Service, Volume 1, July 2, 1987).

U.S. DOE, "The Environmental Survey, Appendix I," DOE/EH-0053, Office of the Assistant Secretary—Environmental, Safety, and Health and Office of Environmental Audit, U.S. Department of Energy, Washington, D.C., 1987.

U.S. DOE, "Soil Sampling Reference Field Methods,", U.S. Fish and Wildlife Service Refuge Contaminant Monitoring Operations Manual, prepared for the U.S. Dept. of Energy, Idaho Falls Operations Office, November 28, 1988.

U.S. EPA. "Sampling and Analysis Procedures for Screening Industrial Effluents for Priority Pollutants," EPA, EMSL, Cincinnati, Ohio, 45268, March 1977, Revised April 1977.

U.S. EPA. "Handbook for Sampling and Sample Preservation of Water and Wastewater," EPA 600/4-82-029, U.S. EPA, Washington, D.C., 1982.

U.S. EPA. "TSCA Good Laboratory Practice Standards," Federal Register, Vol. 48, No. 230, 40 CFR Part 792, November 29, 1983.

U.S. EPA. "FIFRA Good Laboratory Practice Standard," Federal Register, Vol 48, No. 230, 11/29/83, 40 CFR PART 160, 1983.

U.S. EPA. Volume II, Available Sampling Methods," Prepared by GCA Corporation, Environmental Monitoring Systems Laboratory, Report EPA/600/4-83-040, Las Vegas, NV, March 1983.

U.S. EPA. "Interim Guidelines and Specifications for Preparing Quality Assurance Project Plans," QAMS-005/80, EPA-600/4-83-004. Office of Monitoring and Quality Assurance, ORD, U.S. EPA, Washington, DC 20460, Feb. 1984.

U.S. EPA. Federal Register, Part VIII, EPA, "Guidelines Establishing Test Procedures for the Analysis of Pollutants under the Clean Water Act: Final Rule and Proposed Rule," 40 CFR Part 136, October 26, 1984.

U.S. EPA. "Sediment Sampling Quality Assurance User's Guide," U.S. EPA, EMSL, Las Vegas, EP600/4-85-048, July 1985.

U.S. EPA. "Data Quality Objectives Development Guidance for Remedial Investigation/ Feasibility Activities," Document No. 9355.0-7A, EPA/OEER/OSWER, May 9, 1986.

U.S. EPA. "Test Methods for Evaluating Solid Wastes, Volume II: Field Manual, Physical/Chemical Methods, SW 846," Third Edition, U.S. EPA, November 1986.

U.S. EPA. "Characteristics of Hazardous Waste Sites – A Methods Manual, Clayton, C.A., J.W. Hines and P.D. Elkins, "Detection Limits with Specified Assurance Probabilities," *Anal. Chem.*, 59:2506-2514 (1987).

U.S. EPA. "Guidance on Applying the Data Quality Objects Process for Ambient Air Monitoring Around Superfund Sites (Stages I & II)," EPA-450/4-89-015, Washington, D.C., 1987.

U.S. EPA. "A Compendium of Superfund Operations Methods," Volume 2, U.S. EPA, EPA/540/P87/001b, OSWER, August 1987.

U.S. EPA. "Sampling Procedures and Protocols for the National Sewage Sludge Survey," EPA Industrial Technology Division, Office of Water Regulations and Standards, U.S. EPA, Washington, D.C. 20460, 1988.

U.S. EPA. "First International Symposium on Field Screening Methods for Hazardous Waste Site Investigations," Symposium Proceedings, U.S. EPA, EMSL-Las Vegas (October 11-13, 1988).

U.S. EPA. "Soil Sampling Quality Assurance User's Guide," EPA/600/8-89/046, U.S. EPA, EMSL, Las Vegas, NV, March 1989.

U.S. EPA. "Air/Superfund National Technical Guidance Study Series: Volume I—Application of Air Pathway Analyses for Superfund Activities," EPA 450/1-89-001, Washington, D.C., 1989.

U.S. EPA. "Air/Superfund National Technical Guidance Study Series: Volume II—Estimation of Baseline Air Emissions at Superfund Sites," EPA-450/1-89-002, Washington, D.C., 1989.

U.S. EPA. "Air/Superfund National Technical Guidance Study Series: Volume III—Estimation of Air Emissions from Cleanup Activities at Superfund Sites," EPA-450/1-89-003, Washington, D.C., 1989.

U.S. EPA. "Air/Superfund National Technical Guidance Study Series: Volume IV—Procedures for Dispersion Modeling and Air Monitoring for Superfund Air Pathway Analysis," EPA-450/1-89-004, Washington, D.C., 1989.

U.S. EPA. "Methods for Evaluating the Attainment of Cleanup Standards. Volume 1, Soils and Solids Media," EPA 230/02-89-042, Washington, D.C., 1989.

U.S. EPA. "Test Methods for Evaluating Solid Waste" (SW 846). Third Edition, U.S. Environmental Protection Agency, Office of Solid Waste, Washington, D.C., 20460, GPO 955-001-00000-1, 1990.

Index